Life of the Past

James O. Farlow, editor

Rhinoceros Giants

The Paleobiology of Indricotheres

Donald R. Prothero

Indiana University Press

Bloomington and Indianapolis

This book is a publication of

Indiana University Press
601 North Morton Street
Bloomington, Indiana 47404-3797 USA

iupress.indiana.edu

Telephone orders   800-842-6796
Fax orders   812-855-7931

The paper used in this publication meets the minimum requirements of the American National Standard for Information Sciences—Permanence of Paper for Printed Library Materials, ANSI Z39.48-1992.

Manufactured in the United States of America

Library of Congress Cataloging-in-Publication Data

Prothero, Donald R.
  Rhinoceros giants : the paleobiology of Indricotheres / Donald R. Prothero.
    pages cm. -- (Life of the past)
  Includes bibliographical references and index.
  ISBN 978-0-253-00819-0 (cloth : alk. paper) -- ISBN 978-0-253-00826-8 (eb) 1. Indricotherium--Asia, Central. 2. Paleo-biology--Asia, Central. 3. Paleontology--Eocene. I. Title.
  QE882.U6P76 2013
  569'.66--dc23
                              2012036059
        1 2 3 4 5   17 16 15 14 13

This book is dedicated to the memory of
Dr. Malcolm C. McKenna
and
Dr. Richard H. Tedford
for all their contributions to our understanding
of the evolutionary history of mammals.

Frontispiece. The life-sized reconstruction of Paraceratherium, *here shown on display at Morrill Hall at the University of Nebraska, Lincoln (the reconstruction now resides in Gering, Nebraska). To the right are modern African elephants for scale, and in the center is a reconstruction of the running rhino* Hyracodon, *from which indricotheres evolved. (Photo courtesy University of Nebraska State Museum.)*

Behold now the behemoth that I have made with you; he eats grass like cattle. Behold now his strength is in his loins and his power is in the navel of his belly. His tail sways like a cedar; the sinews of his thighs are knit together. His limbs are as strong as copper, his bones as a load of iron.

—Job 40:15–18

CONTENTS

Preface

This book is the culmination of over thirty-five years' worth of research on fossil rhinoceroses, beginning with my first introduction to the Frick and American Museum collections in 1976. I thank Dr. Earl Manning for introducing me to the fossil rhino collections at the American Museum and Dr. Michael O. Woodburne and the late Drs. Malcolm C. McKenna and Richard H. Tedford for all they have taught me over the years. I thank my colleagues Drs. Spencer Lucas, Pierre-Olivier Antoine, Mikael Fortelius, Kurt Heissig, Claude Guérin, and Deng Tao for all their help and efforts in understanding rhinoceros evolution and Drs. Brian Kraatz and Jonathan Geisler for their new insights into Gobi stratigraphy.

The idea for this book emerged from discussions with Dr. James Farlow. I thank Bob Sloan, Angela Burton, Mary Blizzard, and Michelle Sybert at Indiana University Press for all their help in producing the book. I thank Carl Buell for his gorgeous cover art. I thank many colleagues for lending me images; they are acknowledged in the appropriate places throughout the book. I thank Pierre-Oliver Antoine, Mikael Fortelius, James Farlow, Spencer Lucas, and Juha Saari-nen for their helpful reviews of the manuscript. The author designed and laid out the entire book in QuarkXpress 9.3.1 software.

Finally, I thank my amazing wife, Dr. Teresa LeVelle, and my wonderful sons, Erik, Zachary, and Gabriel, for their love and support during the writing of this book on my sabbatical in 2011.

Donald R. Prothero
La Crescenta, California
August 2012

Rhinoceros Giants

Figure 1.1. The American Museum Mongolian expedition, with its Dodge cars and hundreds of camels, near the Flaming Cliffs of Mongolia. (From Andrews, 1932, Plate LV.)

1

Quicksand!

"The New Conquest of Central Asia"

In 1922, the American Museum of Natural History in New York City sponsored one of the most ambitious scientific expeditions ever attempted. Led by the legendary explorer Roy Chapman Andrews (1884–1960), the expedition traveled to China and Mongolia with a huge caravan of seventy-five camels (each carrying 180 kg or 400 pounds of gasoline and other supplies), three Dodge touring cars and two Fulton trucks, and a large party of scientists, guides, and helpers (Fig.1.1). The party included not only Andrews, but also paleontologist Walter Granger (1872–1941), a veteran of many fossil-hunting expeditions in the U.S. and elsewhere, who had prior experience hunting fossils in China. There were also two geologists (Charles P. Berkey and Frederick K. Morris) and many other assistants to drive the trucks and cars and camels, cook the food and set up the camp, and act as guides and interpreters.

The expedition was sent by famous paleontologist and American Museum Director Henry Fairfield Osborn (1857–1935) to find important fossils from Central Asia. Osborn believed that Asia was the center of origin of most mammal groups, including humans, and could contain the legendary "Missing Link" that was long predicted by biologists and paleontologists. Osborn used this argument not only to authorize the expedition, but also to raise funds from his many rich friends who were donors or trustees of the Museum. Osborn told Andrews, "The fossils are there. I know they are. Go and find them."

Andrews provided a colorful and detailed account of all the expeditions in his massive volume with a very un-politically correct imperialist title, *The New Conquest of Central Asia*. One of the most incredible finds of all occurred in the third field season (1925), as described by Andrews (1932, pp. 279–280):

The credit for the most interesting discovery at Loh belongs to one of our Chinese collectors, Liu Hsi-ku. His sharp eyes caught the glint of a white bone in the red sediment of the steep hillside. He dug a little and then reported to Granger who completed the excavation. He was amazed to find the foot and lower leg of a *Baluchitherium* STANDING UP-RIGHT, just as if the animal had carelessly left it behind when he took another stride. Fossils are so seldom found in this position that Granger sat down to think out the why and wherefore. There was only one possible solution. Quicksand! It was the right hind limb that Liu had found; therefore, the right front leg must be farther down the slope. He took the direction of the foot, measured off about nine feet, and began to dig. Sure enough, there it was, a huge bone, like the trunk of a fossil tree, also standing erect. It was not difficult to find the two limbs of the other side, for what had happened was obvious. When all four legs were excavated, each one in a separate pit, the effect was extraordinary [Fig. 1.2]. I went up with Granger and sat down upon a hilltop to drift in fancy back to those far days when the tragedy had been enacted. To one who could read the language, the story was plainly told by the great stumps. Probably the beast had come to drink from a pool of water covering the treacherous quicksand. Suddenly it began to sink. The position of the leg bones showed that it had settled slightly back upon its haunches, struggling desperately to free itself from the gripping sands. It must have sunk rapidly, struggling to the end, dying only when the choking sediment filled its nose and throat. If it had been partly buried and died of starvation, the body would have fallen on its side. If we could have found the entire skeleton standing erect, there in its tomb, it would have been a specimen for all the world to marvel at.

I said to Granger, "Walter, what do you mean by finding only the legs? Why don't you produce the rest?" "Don't blame me," he answered, "it is all your fault. If you had brought us here thirty-five thousand years earlier, before that hill weathered away, I would have the whole skeleton for you!" True enough, we had missed our opportunity by just about that margin. As the entombing sediment was eroded away, the bones were worn off bit by bit and now lay scatted on the valley floor in a thousand useless fragments. There must have been great numbers of baluchitheres in Mongolia during Oligocene times, for we were finding bones and fragments wherever there were fossiliferous strata of that age.

Although Andrews' storytelling skills are vivid, his account of the quicksand is a bit too much like the Hollywood movie version, rather than one based on reality. Fake quicksand in movies sucks the victim down in minutes until he or she is completely submerged. Real quicksand is a slurry of sand and water that remains fairly firm and solid until you disturb it. Then the water between the sand

Figure 1.2. Four indricothere limbs, standing vertically just as they were buried in quicksand. Walter Granger works in the background. (Image number 285735; courtesy American Museum of Natural History Library.)

grains is mobilized, and it becomes what is known as a *fluidized sedimentary flow*. The disturbed water from the pores between the sand grains pushes the sand grains apart so they flow freely, and you can sink down into the slurry. When you stop moving and disturbing the grains, the water also stops moving, and the entire mixture solidifies like concrete—unless you thrash around again and liquefy the mixture. Because a mixture of sand and water is actually denser than water alone, any body (human or animal) will float until it reaches its point of neutral buoyancy. Thus, it is impossible to sink into quicksand above your head. Even if you thrash around in a panic and keep it liquefied so you sink as low as possible, you will still float at the top even higher than you would float in water alone.

The real problem with quicksand is that it is sticky and holds your limbs down so you can't get out easily, and the more you struggle, the deeper you sink until finally you are floating at your point of neutral buoyancy. People or animals trapped in quicksand do not die because they sink down below their head, but because they become exhausted and thirsty since they cannot pull themselves out and remain trapped until they die. If you grab a branch or a rope or any other firm anchor outside the quicksand, you can pull yourself out quite easily—but most creatures trapped in quicksand have no way to pull themselves out. If you should ever become trapped yourself, the best advice is to try to lie flat in it as you would when you float in water and pull yourself out with a stick or rope or any other form of anchoring.

In the case of this trapped indricothere rhino, it probably was mired down to its legs (as they found it), but the rest of the body would not have sank much deeper. Contrary to what Andrews wrote, it would not have toppled sideways if it were half-buried, since the quicksand was thick and stiff enough to trap its legs in an upright pose without allowing the body to lean sideways, let alone escape from it. Instead, with its legs trapped, the indricothere died either of starvation or by being eaten or scavenged by predators taking advantage of the helpless creature. Once it became trapped and could no longer struggle to liquefy the sand, its exposed upper body was easy pickings for predators and scavengers, which is why none of its other bones were there.

Quicksand can be tricky. I've seen a caravan of four-wheel-drive trucks reach a sandy wash where the first vehicle drove across easily. However, its weight disturbed the water and fluidized the saturated sand on the creek bottom, so that when the second truck rolled over it, the sand was mobilized, and the second vehicle sank in to its axles. It became really stuck, requiring trucks on both sides with tow cables and winches to pull it out—*after* everyone had dug out its wheels with a shovel.

The Real "Indiana Jones"?

Roy Chapman Andrews himself (Fig. 1.3) was a flamboyant and colorful character. One of the last classical "scientific explorers" who was not a scientist with a doctorate in a particular specialty, Andrews' gift was in raising funds, in leading and organizing trips, and in conveying the excitement of his many exploits to the general public through his popular books and lectures (which further aided in fund-raising). He was a bold and fearless leader. There are several instances of Andrews scaring off Mongolian bandits by shooting first before they could draw their weapons (and occasionally using guns to intimidate corrupt border guards or greedy officials). In one incident, he charged the bandits with his car, shooting as he approached, causing the bandits' horses to spook, and forcing them to flee. Many people consider him the model of the "Indiana Jones" character played in the popular movie series by actor Harrison Ford, although neither George Lucas, nor Stephen Spielberg, nor anyone else connected with the films has ever confirmed this.

Born in Beloit, Wisconsin, in 1884, Andrews taught himself marksmanship and taxidermy and earned a degree from Beloit College (paid for by his earnings from taxidermy). When he talked his way into the American Museum Director's office and asked to work as a taxidermist, he was told there were no openings, so he started as a janitor. While mopping floors, he also earned his master's degree in mammalogy at Columbia University. In 1909 and 1910, he sailed on the *U.S.S.*

Figure 1.3. Roy Chapman Andrews shown in his typical field attire, with campaign hat and pistol on his hip. (From Wikimedia Commons.)

Albatross in the East Indies, where he collected lizards and snakes and studied mammals. In 1913, he was on the crew of the schooner *Adventuress* in the Arctic, where they hoped to obtain a bowhead whale specimen. In that effort they failed, but he filmed some of the best footage of seals ever captured. In 1916 and 1917, Andrews led the American Museum's Asiatic Zoological Expedition through western and southern Yunnan Province in China, where he collected many specimens and developed valuable skills and contacts in Asia.

By 1920, he was planning the first of several American Museum expeditions to Mongolia. The first was completed in 1922, merely as a short exploratory trip to find out if there were fossils at all. The expeditions were so successful that there were trips in 1923, 1925, 1928, 1929, and 1930. Almost immediately after they arrived in Mongolia, they were finding fossils, and by their second expedition in 1923, they were finding spectacular dinosaur bones and the first known dinosaur eggs. This made the expeditions world famous and helped raise funding

A   B

Figure 1.4.A. Henry Fairfield Osborn. B. Walter Granger. (Courtesy American Museum of Natural History.)

for three additional expeditions. They found not only dinosaurs, but also the first good specimens of tiny mammals from the Age of Dinosaurs.

But by the time of the last expedition, the political situation in Mongolia had deteriorated so badly that no further expeditions were possible. To make matters worse, Andrews and the American Museum had blundered when they auctioned off one of the Mongolian dinosaur eggs as a fundraising gimmick in the winter of 1923–1924. The highest bidder, Col. Austin Colgate, founder of Colgate University, donated it to the Colgate College museum. The auction made the Mongolians angry that their fossils had been plundered, were taken from the country, and were now being sold off at a great price.

By the 1930s, Andrews' ability to mount great expeditions to Asia had ended. The Great Depression made it impossible to raise funds to mount another trip, since many of the formerly rich Museum donors had lost their fortunes in 1929 and 1930, and some of the Museum's investments had become nearly worthless as well. By 1932, the Museum was so strapped for funds that it canceled all fieldwork entirely and cut its staff to the bone. In addition, the tensions between China and Japan were rising as the Japanese prepared to invade China and other parts of Asia.

Andrews spent much of his time in the 1930s writing books about his exploits, being designated one of the first "Honorary Boy Scouts," and serving as President of the Explorers' Club (1931–1934). In 1935, Andrews was appointed Museum Director, but he was unable to do much to help the Museum during the depths of the Depression. Despite his great skills in organizing expeditions and raising money for them, he proved to be so inept at running the Museum that in

1941 the Trustees replaced him. It took two more directors, plus the end of the Depression, for the Museum to recover its strength. Meanwhile, Andrews retired to Carmel-by-the-Sea, where he lived out the rest of his life writing popular books until his death in 1960 at the age of 76.

Osborn and Granger

The other two main figures (and interesting characters) in the American Museum's Central Asiatic Expeditions of the 1920s were Walter Granger (Figs. 1.2, 1.4B) and Henry Fairfield Osborn (Fig. 1.4A). Granger was the main field paleontologist on all the expeditions, and there could not have been a more competent person assigned to the task. Born in Vermont in 1872, he was one of five children of Civil War veteran and insurance agent Charles H. Granger. Like Andrews, he developed an early talent for taxidermy. In 1890, when he was only 17 he got a job doing taxidermy at the American Museum. Within a few years, he was on field expeditions to the American West searching for vertebrate fossils with the American Museum paleontologists. After two field seasons (1894 and 1895), his fossil-hunting talents were apparent, and he was transferred to the Department of Vertebrate Paleontology in 1896. Granger discovered the legendary Bone Cabin Quarry near Laramie, Wyoming, in 1897, which he worked for the next eight field seasons. The site yielded thousands of bones representing sixty-four species of dinosaurs, including the mounted skeletons of the big *Apatosaurus* ("*Brontosaurus*") and *Stegosaurus* currently on display at the Museum today.

After working Bone Cabin Quarry from 1897 to 1906, Granger accompanied Osborn on an expedition to the Eocene-Oligocene Fayûm beds of Egypt in 1907. This was the first American Museum fossil trip outside North America. They made many important discoveries that complemented the discoveries of British Museum paleontologist C. W. Andrews just a few years earlier (see Chapter 2). By this point, Granger was promoted to Assistant Curator and had a flexible schedule that allowed him at least five months in the field every year to find more fossils, while he continued to write two or three scientific papers each year as well. In 1921, he began explorations in China (see Chapter 2), which eventually led to the discovery of "Peking man" in Zhoukoudian caves near Beijing. This work in China laid the foundation for the negotiations to go through China to Mongolia that allowed the 1922 American Museum expedition to succeed. His exploits in Mongolia are discussed later in this chapter. After he finished the expeditions, he was promoted to Curator of Fossil Mammals, which allowed him to continue to work on his many amazing discoveries until his death at age 68 of heart failure. Entirely self-taught, Granger never earned a formal academic degree, though this was rectified in 1932 when he received an honorary doctorate from Middlebury College.

In contrast to the modest and unassuming Walter Granger, who made huge collections and did important research but never promoted himself, there was Henry Fairfield Osborn (Fig. 1.4A). Osborn was born in 1857 in Fairfield, Connecticut, to a wealthy family that was among the elite of New York City. His father, William Henry Osborn, parlayed family wealth and property into a career that eventually made him President of the Illinois Central Railroad, where he built a great fortune. Like many other rich kids, young Henry went to an elite Ivy League school (Princeton), where he fell in love with fossil hunting along with his fellow student William Berryman Scott (who also had a long distinguished career in vertebrate paleontology and collaborated with Osborn many times over the years). In 1877 three ambitious young undergrads (Osborn, Scott, and Francis Speir, Jr.) decided to test their mettle and organized their own "Princeton Scientific Expedition" out in the still unexplored badlands of the West. This was no typical undergraduate field trip, since the area was still overrun by hostile Cheyenne and Lakotas, just one year after the slaughter of Custer's troops at Little Bighorn. A photograph taken of the three of them shows them armed to the teeth. Yet these three brave (or foolhardy) students made a number of important discoveries and returned safely with their scalps intact.

After graduating from Princeton, Osborn traveled and studied in Europe to learn from the leading scientists of the time in Germany, England, and France, and even got to meet Charles Darwin and to shake his hand. (When I was a grad student at the American Museum-Columbia program from 1976 to 1982, I got to shake hands with Edwin Colbert, who had shaken Osborn's hand many times, so I'm only three degrees of separation from Darwin.) After his return from Europe in 1883, Osborn taught at Princeton. In 1891 he was hired jointly by the American Museum and Columbia University, where he set up the program that trained some of the foremost vertebrate paleontologists in the world for a full century. In 1908, he became the American Museum President, serving until 1933. More than anyone before or since, Osborn was responsible for building up the museum's exhibits and collections and buildings and for making it one of the foremost natural history museums in the world. During this time, Osborn was also Chair of the Department of Vertebrate Paleontology and made that collection and program the largest and most important in the United States.

Although Osborn was a superb administrator and a genius at fund-raising among his rich friends, today most paleontologists regard him as a less-than-stellar scientist. Osborn's notions of paleontology and systematics were outdated and idiosyncratic, even for his time. His voluminous works were full of taxonomic oversplitting, and he created hundreds of invalid species. These huge scientific monographs on rhinos (1900), horses (1918), brontotheres (2 volumes of over 1,000 pages published in 1929), and the proboscideans (1936) were so full of errors and ideas that are unacceptable by modern standards that they have hindered research progress for decades. Osborn had bizarre notions of evolution, some of which were influenced by his own background as a product of wealth and privilege. He was a follower of eugenics and racist anthropology and advocated policies toward non-white races that are offensive today. He believed in the superiority of the western European-American Caucasian race and often made scientific mistakes due to his racist biases (such as endorsing and writing a whole book about the Piltdown skull, later found to be a forgery). Osborn mistook a worn peccary tooth from Nebraska for an anthropoid primate, later known as "Nebraska Man"; this mistake was later quietly corrected by Osborn's colleague William King Gregory. Osborn believed that evolution proceeded in straight lines ("orthogenesis") without the control of natural selection, often evolving into "racial senescence," or forms with what he thought were maladaptive features. These included examples like the huge antlers of the "Irish elk" or the teeth of a sabertooth (all of which have been restudied and shown to be adaptive after all). He thought that evolution strove to produce the "best," most fit lineages ("aristogenesis") and weeded out the inferior races, consistent with his own position at the top of the American aristocracy. Even Osborn's own colleagues at the American Museum, like the much more competent William Diller Matthew and William King Gregory, found his ideas unacceptable, but they kept their disagreements out of the press.

The stories about Osborn's egotistical behavior are legendary. Rainger (1991) describes him as a powerful, well-connected American aristocrat, running the department and the museum as a strict hierarchy, where he provided the administration and financial means. His subordinates (many of whom were more competent paleontologists than Osborn) did all the grunt work (field work, preparation, curation, illustration, writing, and description), for which Osborn took complete or at least partial credit. He had a retinue of secretaries and assistants who attended to his every beck and call, and they even had separate quarters outside his mansion in Garrison, New York, so they could do his bidding when he was at home. At social events, only his top scientific assistants and fellow curators could dine with Osborn and his family; the "lesser staff" were consigned to a different room. George Gaylord Simpson, the legendary mammalian paleontologist who overlapped with Osborn's later career, describes an incident when he tried to apply an ink blotter to Osborn's autograph from a flowing fountain pen. Osborn stopped him and said, "Never blot the signature of a great man."

Nevertheless, for all his flaws, Osborn accomplished an amazing amount in his lifetime, not only building the American Museum and organizing and funding expeditions to build the biggest collections in the world, but also describing, naming, and publishing hundreds of famous fossils, from *Tyrannosaurus rex, Velociraptor, Pentaceratops,* to hundreds more dinosaurs and fossil mammals. During the last years of his life, Osborn was in failing health, and the Museum was struggling due to the Great Depression. He stepped down as President in 1933 and died in 1935 at age 78, before he could see his final great work, the huge Proboscidea monograph, published. For better or for worse, every vertebrate paleontologist owes some of the prominence of our field to his work, and most of us have to deal with his published ideas many times in our own careers.

Into the Gobi Desert

The "biggest scientific expedition ever to leave the United States" passed through a gate in the Great Wall of China on April 21, 1922, and headed into Mongolia. Aided by their cars, they traveled 426 km (265 miles) in just four days, covering ground much more efficiently than any previous expeditions mounted on horse or camel. Over the course of the many expeditions, they faced bandits, blinding sandstorms, blistering heat and freezing cold, and lots of dangerous vipers, but the trips went off with relatively few problems considering the incredible discoveries that were made.

Often overlooked in all the excitement over dinosaur skeletons and eggs were many spectacularly large fossil mammals, including huge shovel-jawed mastodonts, a giant predatory mesonychid hoofed mammal, named *Andrewsarchus* in honor of their leader, and the very first good fossil record of the evolution of mammals in the Paleocene, Eocene, and Oligocene epochs of Asia. But the most spectacular of all were the gigantic hornless rhinos known as indricotheres. Even before Andrews had set up camp on the first days of the 1922 expedition, they found fossils. These were fragmentary specimens found near Iren Dabasu by one of their geologists, Charles Berkey, in April 1922, when he was scouting the region and mapping the geology. The group dropped off Walter Granger at the outcrop while they began to set up camp. According to Andrews:

> We were hardly settled before Granger's car roared into camp. The men were obviously excited when I went out to meet them. No one said a word. Granger's eyes were shining and he was puffing violently at his pipe. Silently he dug into his pockets and produced a handful of bone fragments; out of his shirt came a rhinoceros tooth, and the various folds of his upper garment yielded other fossils. He held out his hand: "Well, Roy, we've done it. The stuff is here. We picked up fifty pounds of bone in an hour."

By June 26, the expedition was near Mount Uskuk at what they called "Wild Ass Camp," so named because wild asses were common in the region. As Andrews (1932, pp. 103–104) described it:

> The day after our arrival at "Wild Ass Camp," I found my first important fossil. We could prospect within a dozen yards of our tents, which were on the edge of a red and white wash. In the morning [expedition cinematographer James B.] Shackelford found a beautifully preserved foot bone of an unidentified rhinocerotid type, in the bottom of a deep ravine. I was stimulated to better his discovery. After setting a line of traps in the river bottom near the well, I was wandering slowly along the sides of the ravine looking for fossils. Fragments were abundant

but none were identifiable or worth keeping. Suddenly my eyes marked a peculiar discoloration in the olive-green upper strata; then I could see bits of white which looked like crumbled teeth. Scratching away at the soft claylike earth, I exposed the grinding surface of some large teeth, and realized that it was something interesting. There was an almost irresistible temptation to dig further and see what lay below, but I knew that if I did the wrath of the palaeontologist would descend upon my head. The teeth were literally in powder, and fell apart in a thousand tiny fragments when the supportive earth was removed. Because of its bad state of preservation, Granger was doubtful at first whether it would be worthwhile removing the specimen, but eventually he decided to make the attempt.

It took four days of intermittent work to get it out, and no one but a master of the technique like Walter Granger could have accomplished it at all. By means of a fine camel's-hair brush, he removed the sand almost grain by grain, wetting the teeth with gum Arabic as each minute section was exposed and stippling Japanese rice paper in the crevices. When the gum and paper dried, the dust-like particles of enamel were so cemented that it was safe to expose a still larger surface. Then Granger soaked strips of burlap in flour paste and bandaged the fossil as though it was a broken limb; after a day of sun this formed a hard shell in which the specimen was safe.

As the work progressed it became evident that much more than a set of teeth lay buried in the hill; one side of the palate was exposed, then the jugal arch which forms the cheek, and finally the anterior part of the skull with a pair of extraordinarily long decurved nasal bones. The teeth showed that the animal was a rhinoceros of a kind that none of us knew. Subsequently it was studied by Professor Henry Fairfield Osborn, 1924, who named it *Baluchitherium mongoliense*.

Then the expedition went to the badlands of the Loh area and the Tsagan Nor Basin. When Andrews and Granger reached these Oligocene beds (called the Hsanda Gol and the Loh formations—see Chapter 3), they found the ground littered with huge bones of a gigantic rhinoceros, including a huge skull and jaws and limb bones of elephantine proportions (Fig. 1.5). Andrews (1932, Chapter XIII) described finding isolated bones that the Mongols had said "were as large as a man's body" and realizing that they were not exaggerating. As Andrews (1932, pp. 142–143) described it:

The next evening, while we were silent with awe witnessing the most glorious sunset that I have ever seen, a black car slipped into camp from the north, carrying Granger and Shackelford. Their work at Loh was finished and they came down with all their gear and specimens.

Figure 1.5. Shackelford (left) and Granger (center) and a Chinese assistant working on huge indricothere limbs. (From Andrews, 1932, Plate VI.)

Their faces were radiant with suppressed news and when the celestial display was ended and the soft evening light had enshrouded the rugged outlines of Baga Bogdo in a robe of delicate purple, they produced a prize exhibit—newly discovered bones of a giant *Baluchitherium*! These had been found after Granger and Shackelford had broken camp and were on their way to the lake. So it happened all through the summer! We had only to pack and make ready to leave a fossil bed to discover some priceless specimen!

Granger and Shackelford had decided to walk through a still uninspected pocket in the badlands and to have Wang, their Chinese chauffeur, drive the car ahead to a promontory two miles to the south. After a little while, Wang decided to do some prospecting on his own account. Almost immediately he discovered a huge bone in the bottom of a gully that emptied into a larger ravine. It was the end of a humerus of a *Baluchitherium*, and other parts were visible, partially embedded in the red earth.

We had impressed upon all our natives the importance of leaving a bone in situ, and Wang waited until Granger arrived. Then they found one side of a lower jaw and other fragments of the skeleton. The bones were so hard and free from matrix that they could be easily removed, and Granger brought everything to camp at once.

We talked far into the night, and I awoke the next morning field with a desire to find more of the *Baluchitherium*. Granger could not leave camp, as he was packing fossils to go with the camels which we wished to start eastward as soon as possible. Therefore, Shackelford, Wang and I drove back to Loh to give the gully a more careful inspection. Shackelford and Wang began to dig a trench where the jaw had been found while I prospected the sides of a small dividing ridge. Arriving at the crest of the ridge, I looked down the other side and instantly saw fragments of bone half covered with loose sand in the bottom of the wash. I knew that the bone fragments were portions of a *Baluchitherium* because of the color and leaped down the slope with a shout. Shackelford and Wang came on the run and in short time we had unearthed several large fragments; then our fingers struck the end of huge block in which a loose tooth was embedded. Evidently we had discovered a skull. I knew that it was time to stop, and we returned to camp with our spoils, for I was too excited to do further prospecting that day. We burst in on the camp as the men were having tea. The next day the party returned to Loh to begin the work of removing the huge bones.

The skeleton evidently had been buried in a spot that later became the summit of a ridge between two gullies. As the sediments weathered away, part of the skeleton rolled down one side of the ridge; that was what Wang found on the first day. The rest had slipped down the other side into the main wash, where I discovered it. We followed the ravine down to the mouth, rescuing a few fragments that had been carried away by rain action, and Shackelford discovered several important sections of the skull at least three hundred yards out on the plain.

We sifted every square inch of sand, and at the end of four days which Granger had required to remove the skull, we felt certain that nothing had been overlooked. The huge block was strengthened and packed in two sections in camel boxes preparatory to its long journey to New York.

As Preston (1986, pp. 99–101) described it, the specimen eventually proved to be one of the most complete indricotheres ever found, and the only one from Mongolia with a good skull (Fig. 6.1). It was hauled in two huge blocks across Mongolia and China, protected from bandits, and then shipped halfway around the world to New York. When Osborn's assistants opened the crates and prepared the specimen, Osborn realized the extraordinary quality of the specimen. In his description (1923), he wrote that he considered it one of the great moments in the history of paleontology.

The 1928 expedition to Mongolia focused on the Baron Sog Formation at a location called Holy Mesa near Urtyn Obo. Andrews (1932, p. 392) tells of how indricothere bones were so abundant that he could spot their white color from a distance using binoculars, and specimen after specimen was recovered. His final account of a major discovery of indricotheres in Mongolia is from June 25, 1928 (Andrews, 1932, pp. 393–395):

The next day Shackelford made an astonishing discovery. He came into the mess tent for tiffin and casually remarked that he had found a "bone." Rather too casually, I thought. I was sure that the half had not been told. After suitable encouragement, he admitted that it was a large bone—a very large bone. Only the end of it was projecting form a hill slope, but that end was as big as his body. There was a roar from the table at that, for Shack's body is far from thin. He is not exactly globular, but he certainly is fat. "Don't believe me then," quoth Shack, "but I'll show you."

And show us he did. Walter Granger, Albert Thomson, Shackelford and I went there in a car, for the place was two miles from camp. It proved to be a gray slope, which dropped off abruptly into a deep ravine. Ten feet down the side lay a great white ball. Until I examined it I would not believe that it was bone, for it actually was as thick as Shackelford's body. A little brushing off of yellow sand showed it to be the head of a humerus, or upper arm bone. More brushing exposed its entire length and brought to light the end of another massive shaft, which ran deep into the hillside.

All of us stared in amazement. It is not easy to ruffle the calm of Granger and Thomson. They have been at it too many years and have dug up too many strange beasts. But they got a real surprise when they saw those bones. As for me, I was too impressed even to talk. The size was almost terrifying. The humerus which Shackelford had found was as thick as a man's body and three and one half feet long. The second giant shaft proved to be the radius. It was nearly five feet long, and so heavy that two of us could hardly lift it. In order to thoroughly to prospect the deposit, the side of the hill must be removed; it might reveal an entire skeleton.

So at seven o'clock the next morning half the men of the Expedition began shoveling energetically at the coarse yellow sand. The bones were so hard and big that there was little risk of breakage; therefore, Granger allowed me to work around them with a curved awl and a whisk broom. My pickaxe methods do get quick results, but they are a bit rough on the specimens, I must admit. In the language of the Expedition, when a fossil is broken beyond repair it has been given the "R.C.A."

Before the massive radius lay bare for its entire length, Granger discovered another from the opposite side; also two enormous ribs. Just

Figure 1.6. The entire crew (Granger, Shackelford, Thomson, and Andrews) at the indricothere quarry. (From Andrews, 1932, plate XC.)

behind them, farther in the hillside, my brush exposed a corner of a flat bone; then a huge tooth, nearly as large as an apple, came into view. That gave all of us a thrill, for a skull with teeth meant that we could positively identify the specimen, but it proved to be only a jaw, and the left side was gone. The doctor next uncovered the middle metatarsal of one foot.

Then we paused to have a look at things [Fig. 1.6]. The shoveling squad had removed fifteen feet of hillside, leaving a flat bench where the bones lay exposed. They were all on the same level, close together and the ends pointed in the same direction. It was obvious that the deposition had taken place in the bed of a swift stream, flowing north. The cross-bedding of the yellow gravel and the position of the bones told the story. The animal had died in the stream, the flesh decomposed and the skeleton disarticulated. The smaller parts had been carried on by the water; doubtless many had been broken by pounding against rocks. The massive limb bones had been left where the beast died. They were too heavy even for a torrent to move more than a few feet. It was useless to dig farther into the hill, for we were rapidly getting out of the streambed. Only excavations along the watercourse, northward, would yield results, but unfortunately a deep ravine had cut through it in that direction, and the ancient bed was gone. Extensive excavations farther into the hillside yielded no more bones, and we had to admit, reluctantly, that there was no hope of finding more at that spot.

Since their discovery on the American Museum Mongolian expeditions of the 1920s, the fossils known as *Baluchitherium* or *Indricotherium* or *Paraceratherium* have amazed and baffled both scientists and non-specialists alike. How did these enormous creatures live? What did they eat? What kind of environment did they live in? Where did they come from? What happened to them? These questions, and many more, will be the subject of the remaining chapters.

2

Giant Hunters

Pilgrim's Progress

Before we look further into the life of indricotheres, we need to discuss the places where they have been found and the nature of the fossils discovered so far. This naturally leads into a story of the paleontologists who have taken great risks to travel to remote and dangerous places, from central Asia to Mongolia to China to regions of what is now Pakistan (Fig. 2.1). As we saw with Andrews' account of the American Museum Mongolian expeditions, almost all of these discoveries were made at great risk and after enduring much hardship. The rugged individuals and pioneering paleontologists who made these discoveries are just as colorful and interesting as the extinct creatures they found.

The story of the discovery of indricotheres begins with just such a colorful pioneering paleontologist, Henry Guy Ellcock Pilgrim (Fig. 2.2), whose career was described by Lewis (1944). Born on Christmas Eve 1875 at his family's colonial mansion in Barbados, Guy Pilgrim began his education at Harrison College on the island. Unable to get the education there that he needed to further his career goals, he traveled to England and transferred to University College, London. There he finished his B.S. degree in 1901 and eventually earned his D.Sc. degree in 1908.

However, right after he earned his B.S. degree he began his remarkable career as a geologist and paleontologist in the Middle East and southern Asia. In 1902, he was appointed to a post at the Geological Survey of India, then a British colony. From that position the ambitious young man traveled widely over southern Asia and the Middle East, doing geological mapping and reconnaissance for the Crown Colonies and finding numerous fossils. This work was truly groundbreaking because almost no real geologic mapping or research had ever been conducted over this wide region before. Pilgrim was a true pioneer. His travels took him to Arabia, Persia (now Iran), Baluchistan (now part of southern Pakistan),

Figure 2.1. Index map of some of the early indricothere localities in Asia, with their political boundaries as of 1923 (Osborn, 1923a.) 1 is Dera Bugti, Baluchistan, Pakistan; 2 is the location of the first Indricotherium *specimens north of the Aral Sea, Kazakhstan; 3 are the Hsanda Gol and Loh formations, Mongolia; 4 is Iren Dabasu, Mongolia.*

Bhutan in the Himalayas, and the Punjab and Simla Hills in British India. By 1905, he was also appointed as a paleontologist at the Geological Museum of Calcutta, eventually making the post of curator in 1909 after he finished his doctorate. Within a few years he began to publish his travels and research, starting with two memoirs on the geology of the Persian Gulf and the Arabian Peninsula, the first published in 1908 and the second later in 1924. Later, oil geologists used his pioneering mapping to discover the oil wealth of the Persian Gulf, where most of the world's oil is still produced. Pilgrim was the first European to visit Trucial Oman, and the first geologist to explore Bahrain. His discovery of the domed structure there eventually led to the great discoveries of oil in Bahrain and other Persian Gulf countries.

However, his greatest paleontological contribution came from mapping and fossil collecting in the Siwalik Hills of what is now Pakistan, which is considered one of the most complete and fossiliferous continuously exposed sequences of Miocene and Pliocene rocks anywhere in Asia. Earlier paleontologists, such as Hugh Falconer, Proby Cautley, and Richard Lydekker, had already made the first collections there, roughly mapped some of the geology, and described some of these fossils when they reached Britain, but their work was preliminary and in-

Figure 2.2. Guy Pilgrim. (Image courtesy Royal Society of London.)

complete. Pilgrim mapped and collected the area in much greater detail and realized that the Siwalik sequence spanned a great deal of time with many different
successive faunas showing remarkable evolutionary change. Since the 1970s so
much research has been done on dating the Siwalik sequence and studying its
mammals that it is considered the "gold standard" for understanding the last 23
million years of climate and evolution in Asia. According to Lewis (1944), the
legendary paleontologist W. D. Matthew wrote: "The admirable later work of Pilgrim was the first to make clear the distinctions between the successive faunas,
and added very largely to the faunas." Edwin H. Colbert (who worked on Siwalik
mammals himself in his early career before turning to dinosaurs) wrote: "Dr. Pilgrim virtually opened the field in the discovery and study of Lower Siwalik vertebrates." Pilgrim's first publication on the topic was *Preliminary Note on the
Revised Classification of Tertiary Freshwater Deposits of India* (1910) followed
by *The Correlation of the Siwaliks with Mammal Horizons in Europe* (1913).

When World War I broke out, the British Army was busy in the Middle East
defending its colonies and battling the Turkish Army, which was allied with Germany and Austria. This was the same British Army portrayed in the movie
Lawrence of Arabia. In the movie, and in real life, the British needed T. E.
Lawrence to help them unify and forge alliances with the Arabs of the region to
fight the Turks. As one of the most educated and widely traveled British scientists
in southern Asia and the Middle East at the time, Guy Pilgrim also served as a
valuable resource to the Army. Although the details of his service are not fully
disclosed, after the war he was decorated for his war service in Persia and
Mesopotamia (now Iran and Iraq).

After his war service ended in 1919, Pilgrim returned to his posts in India, and in 1920, became Superintendent of the Geological Society in India. For the rest of his career, he rose through the ranks of scientists in colonial India and received numerous honors, including Fellow of the Asiatic Society of Bengal (1925), President of the Geological Section of the Indian Science Congress (1925), honorary staff appointment at the British Museum of Natural History in London (1930), and finally in 1943 he was elected to the most prestigious honor of all, Fellow of the Royal Society of London.

Meanwhile, he continued his research and finished a stream of important publications, including *The Geology of Parts of the Persian Provinces of Fars, Kirman, and Laristan* (1925); *The Structure and Correlation of the Simla Rocks* (1928); *Catalogue of Pontian Carnivora of Europe* (1931) with A. T. Hopwood in the *British Museum Bulletin; Siwalik Antelopes and Oxen in the American Museum of Natural History* (1937), which was published in the *American Museum Bulletin* after a long visit to the U.S. that started in 1932; and his crowning achievement, *Fossil Bovidae of India* (1939). His research on the fossil apes of the Siwaliks is still used by anthropology scholars today, and his research on fossil carnivores and bovids (cattle, goats, and antelopes) is the foundation of all later research in these groups. By the 1940s, he was retired and living in England, where he passed away at his estate in Upton near Didcot, Berkshire, on September 15, 1943, at the ripe old age of 78. As paleontologist G. Edward Lewis of the Smithsonian wrote in his obituary notice for Pilgrim (1944), "no tribute in memory of him would be complete without mention of one of his most important services to science: the encouragement of others in studies of his chosen field. Many younger men were privileged to receive his generous friendship, disinterested advice, and inspiring help, to their own great benefit." As a testimony to his influence, there are several fossil species named after him and a genus of anthracobunid (distantly related to early proboscideans) called *Pilgrimella*.

Pilgrim is important to our story because he was the first to discover and name indricothere specimens, even though he did not realize it at the time. During his wide-ranging exploration and collecting in India in 1907–1908, he was one of the first to travel to and collect fossils in the Dera Bugti area south of the Siwaliks in what is now a very dangerous tribal region of southern Pakistan. In his 1912 *Memoir of the Geological Survey of India*, he published the first notice of the loose upper and lower teeth and the back of a jaw of an indricotheres. Pilgrim named the specimens *Aceratherium bugtiense,* throwing the new species into a "wastebasket" genus *Aceratherium* that was then used for nearly every hornless rhinoceros fossil of the Oligocene and Miocene in Eurasia and North America. That genus is still valid today, although it is restricted to just a few species that are closely related to the type species. Most of the other species thrown into "*Aceratherium*" because they were hornless are now in other genera.

Figure 2.3. Sir Clive Forster Cooper. (Image from Wikimedia Commons.)

Forster Cooper's Finds

The next phase of indricothere research occurred even before Pilgrim's 1912 publication with the work of Sir Clive Forster Cooper (Fig. 2.3), whose biography was summarized by Watson (1950). Born in the Hampstead area of London on April 3, 1880, he was the son of John Forster Cooper, a solicitor, and the family's lineage could be traced all the way back to at least 1427. (The name "Forster Cooper" is a typical British double-barreled surname, which in most publications is not hyphenated, although in some journals it is.) Forster Cooper went to a private school (Summerfields, Oxford) and then attended Rugby for his secondary education. He enrolled at Trinity College, Cambridge, at the age of 17 in 1897, and studied ("read" in the English convention) zoology, physiology, and geology. At the beginning of his second year, Prof. J. Stanley Gardiner saw him dissecting in the laboratory and invited Forster Cooper to join his research program in the Maldives Islands in the Indian Ocean in 1899. Using a small schooner, they made important natural history collections both on the islands and just offshore in the coral reefs. Gardiner then contracted malaria and had to return to Colombo, Ceylon (now Sri Lanka) for treatment, leaving Forster Cooper in charge of the expedition at the age of 20. Despite his youth and the hazards of collecting around a coral reef, the difficulty in keeping good records on the specimens, and the responsibility of managing a ship with twenty crewmembers, Forster Cooper proved his competence and maturity. When he returned to Cambridge, he finally finished earning his degree in 1901.

Forster Cooper then immediately returned to exploring, serving as the naturalist on the International North Seas Fisheries Commission in 1902–1903, then joining Gardiner again in 1905 on the *HMS. Sealark* voyage exploring the Indian

Ocean (especially the Seychelles). He returned to Cambridge in 1906 to continue his research on the specimens from his two Indian Ocean trips, only to immediately join Dr. Charles W. Andrews of the British Museum of Natural History on his legendary expedition to the Eocene-Oligocene fossil mammal beds of Fayûm, Egypt, in 1907 (Osborn and Granger were also working in the same area). These deposits produced fossils not only of the earliest mastodonts and hyraxes, but also early whales and many other bizarre extinct beasts, such as arsinoitheres. The most famous fossils of the Fayûm, however, are some of the oldest and best-preserved fossil apes and monkeys ever discovered.

His experience collecting fossil mammals in Egypt shifted his research focus from marine biology to mammalian paleontology. He spent a year at the American Museum of Natural History in New York learning about the much better fossil record of North American mammals. Along with studying the Museum's immense collections, he learned from legendary paleontologists like Henry Fairfield Osborn, William King Gregory, William Diller Matthew, and Walter Granger and joined them in their field expeditions to Wyoming. When he returned to Cambridge in 1910, he learned of the new discoveries in India made by Dr. Guy Pilgrim and made arrangements to visit Pilgrim's Dera Bugti localities in Baluchistan (now in southern Pakistan). His first expedition in 1910 was a success, and it was soon followed by an even larger expedition in 1911. This time he found a large bone bed of Pilgrim's "*Aceratherium*" *bugtiense,* and collected a large series of skulls and partial skeletons of the gigantic rhinoceros for the first time.

Even though he had the best possible field training for that time from his work with Granger and Matthew, the fieldwork in Dera Bugti was challenging. The bones were well-preserved but huge and very fragile, so that even with good plaster jackets around them (as he had learned from working with Granger) they were hard to retrieve and take back to civilization. The only transport was on the backs of camels, which can carry a great amount of weight as long as it is balanced on a sling on each side of the camel's back. Forster Cooper found a complete indricothere pelvis, which is larger than that of an elephant. When it had been excavated and jacketed, it was too huge and too heavy for one camel to carry. They tried to suspend it between two camels, but it quickly disintegrated while it was being carried. Nevertheless, Forster Cooper's expedition successfully removed most of the bones, aided not only by his field experience, talent, intelligence, and good sense, but also by his endearing personality that charmed the Baluchis, on whose land he was collecting.

When Forster Cooper returned to Cambridge in 1911, he immediately began removing all the field jackets and doing the delicate preparation of the specimens by himself (a job usually left to expert preparators today). By the end of that year, he had published a paper renaming Pilgrim's "*Aceratherium*" *bugtiense* as *Paraceratherium bugtiense* (*Paraceratherium* means "near the hornless beast"), since it was clearly not a European hornless rhino, but appeared to be a giant relative of this form. In 1913, he named another specimen from the collections *Thaumastotherium* ("wonderful beast" in Greek) *osborni*, which later proved to be an invalid name for *Paraceratherium bugtiense*.

World War I caused an interruption of his paleontological research. As someone with medical training, he was asked to do research on human animal parasites, especially on how to treat malaria, as a contribution to the British war effort. Once he had finished his war service in 1921, he resumed the post of Superintendent of the University Museum of Zoology in Cambridge, a position he had originally accepted in 1914. Although he was busy teaching a number of students (many of whom went on to prominent careers in zoology) and taking care of administrative duties (such as rebuilding the Cambridge University library system), he found some research time to further study his indricotheres. He published two more papers in 1923, one naming *Baluchitherium osborni* as a replacement name for *Thaumastotherium*, a name that was already in use for a hemipteran bug, and a second giving a detailed description of the skull and teeth of his *Paraceratherium* fossils. His final major paper on the topic appeared in 1934, with a summary of nearly everything he had learned from his Dera Bugti collections. He published many other papers on fossil mammals from Asia, including carnivores, chalicotheres (extinct horse-like creature with claws), anthracotheres (primitive relatives of whales and hippos that were once widespread in Eurasia and North America), ruminants, mastodonts, and elephants.

In 1938 at the age of 58, he moved up from his Cambridge post to the Directorship of the British Museum of Natural History, where he made even bigger collections, traveled widely, and bolstered the scientific standing of this august institution. There, he could not only study his own Asian fossils, but eventually tried to sort out the complicated systematics of the Eocene British horse-like material of *Hyracotherium*, Ice Age British elephants, and British Paleozoic fishes. He even introduced a novel approach to understanding the skull bones of fossil lungfish that is still followed today. As Director, his main focus was to modernize the ancient "collector's cabinet" galleries of the British Museum, which contained dusty, poorly lit specimens and faded labels. He wanted to make the Museum into a grand gallery of freestanding fossil skeletons, along the lines of the museums in New York and at Yale.

However, his tenure as British Museum Director came as his health was declining from frequent sinus infections. By 1939, he also had to face the events leading to World War II and the imminent German attack on England. Even before the war started, Forster Cooper made arrangements to ship the most scientifically important specimens (from thousands of specimens in glass jars full of flammable alcohol to stuffed birds and mammals to many tons of fragile fossils) to Tring in Hertfordshire, a small town 30 km north of London that was unlikely to be bombed by the Germans. By the time the war started, the Museum was continually occupied by fire-watchers on patrol to put out any fires caused by bombs, and Forster Cooper himself spent most of his time in the Museum, even sleeping there at night. The Museum was hit several times in the 1940 Battle of Britain and again in 1944, but most of the valuable collections (except for a suite of rare British birds' nests in glass cases) survived unscathed. When the war ended, the Museum was in shambles, with most of its windows blown out, its cases shattered, the roof destroyed, and only a skeleton staff remaining. Forster Cooper began working hard to restore the Museum, but transport and materials and money were in short supply, and he was over 65 years old and in poor health. He died on August 23, 1947, at the age of 67, still working hard as Museum Director, but he did not live to see his beloved Museum restored or all the exhibits redesigned into the modern dynamic form that he long envisioned.

Although Forster Cooper did not publish as much as he might have, he spent much of his life in administrative posts and had his research time and energy and institutional resources diverted by two world wars. His service to British science was recognized by election to the Royal Society in 1936 and a knighthood in 1946. He was also honored by international organizations and made a foreign member of the New York Academy of Sciences. His name lives on not only with the early indricothere first named *Cooperia* by Horace E. Wood II (later changed to *Forstercooperia*), but also with a mastodont named by Osborn, *Trilophodon cooperi* (now *Gomphotherium cooperi*). More importantly, he was the first to recognize that indricotheres were strange and very unusual rhinos that did not belong in the family Rhinocerotidae, even though he had only partial skulls and skeletons on which to base his research.

Borissiak and the Russian Giants

Almost as soon as the Soviet Union formed in 1917, its leaders transformed the scientific organizations that had been under the patronage of the Czar into Soviet Academies that were beholden to the state. There are many stories of how this system was abused by politically connected scientists. One of the most infamous of these was the saga of Trofim Lysenko, a favorite of Stalin, whose false ideas of genetics caused untold millions of Russians to starve when crops failed, even as he led the persecution and exile and murder of Russian geneticists who followed Mendel's principles. In other cases, solid science was being pursued, although often in isolation from scientific developments in the western world.

As we shall see in Chapter 3, in the 1940s and 1950s (and sometimes even today) Soviet (and also Chinese) scientists maintained outdated late nineteenth-century ideas of how to define a species. Meanwhile European and American scientists were beginning to apply modern concepts of biological species and populations to the naming of fossil species and statistical approaches to fossil samples. Due to their fondness for Marxist theory and their ideological preconception that earth history is revolutionary and punctuated by sudden events, Soviet scientists insisted on concepts in stratigraphy (such as the false notion that any rock formation is a time unit as well) that vanished from the thinking of non-Soviet geologists (see Prothero and Schwab, 2012).

The isolation between Soviet and Western scientists was partly due to the language barriers. Few Russians could read English or were allowed to travel outside the Soviet Union, and even fewer Western scientists learned Russian or were allowed to visit Soviet museums. There were also political barriers, since the U.S. refused to recognize the Soviet Union until becoming their allies in World War II. Then after the war ended in 1945, the U.S. was locked in a "Cold War" with the Soviet Union for the next forty-five years. Thus, huge riches of fossil resources were accumulating in Soviet museums, unseen by Western paleontologists, since most were not published or (if they were published) were in obscure hard-to-obtain Russian journals with crummy illustrations and texts entirely in Cyrillic.

My good friend and colleague, the late paleontologist Everett "Ole" Olson, was one of the few paleontologists to make contact and to collaborate with Soviet scientists during the Cold War. He described his first visit to the Paleontological Museum in Moscow in the 1950s (Olson, 1990, p. 99):

When I first visited Moscow, the Paleontological Museum was housed in the left wing of the old Mansion of Orlov dating from the time of Catherine II. The main building was occupied by the Presidium of the Academy of Sciences of the USSR and the coach house wing housed the Paleontological Museum at one end and the Mineralogical Museum at the other. There is a new museum building now, but I miss the old one. It was in the midst of its clutter of bones and display cases that Efremov and *Estemmenosuchus* were juxtaposed during one of my early days studying there. Even at this time, Efremov had a bad heart and was spending much of his time, to the dismay of Professor Orlov [Academician Yuri Orlov, head of the Paleontological Institute at that time] in a dacha of the Academy of Science writing novels. Just inside the front entrance to the Museum, a giant skeleton of a duck-billed dinosaur from the Gobi Desert formed something of an arch over the door of the office of the gracious Director of the Museum, Konstantine Flerov. Beyond in the main hall, a near jungle of cases of skeletons and bones, many from the Cis-Uralian areas and some from the Gobi Desert, rose in ordered disarray. This was what I had come to study. In one case, some 35 skulls of a primitive pareiasaurian reptile were stacked in a pyramidal fashion. In this museum were the treasures of the Russian Permian which, except for the most general facts about them, had essentially been "lost" to the western world until about 1954. Since then, many other scientists have passed through the doors during their own visits. The main barrier, language, still remains, and few publications other than the technical *Paleontological Journal* are regularly translated into English. Year by year circumstances improve, but much of the volume and detail of Soviet day-to-day science remains obscure.

Despite these handicaps, the Soviet paleontological community made many important advances, largely because the former Soviet Union included massive areas of central Asia and Outer Mongolia that were covered by huge fossil-rich badlands. Even before 1916, M. V. Bayarunas of the Geological Museum of the Russian Academy of Sciences conducted expeditions in the Turgai region north of the Aral Sea. Near the Kumbulak Cliffs east of the town of Agypse (Akespe) along Perovsky Bay on the north shore of the Aral Sea, Bayarunas and his expedition recovered a nearly complete skeleton (Fig. 2.4) of a giant rhinoceros from the Upper Oligocene Aral Formation.

When this spectacular set of fossils was brought back to Leningrad, it was the prerogative of the Director of the Paleozoological Institute, Academician Aleksei Alekseeivich Borissiak (1872–1944), to describe the specimens. He published a short note in 1916 that called the specimen *Indricotherium,* after the Indrik beast, a monster in central Asian mythology that could fly above the clouds and caused the earth to tremble when it walked. He did not assign a species name to the material at that time but waited until a later paper in 1923 when he called it *Indricotherium asiaticum.* However, his delay meant that other scientists could assign a species to the material. In 1922, Maria Pavlova named the material *Indricotherium transouralicum,* which became the formal name for these specimens. Borissiak published longer descriptions of the fossils in the 1920s, and the huge skeleton became one of the prize exhibits (and still is an amazing specimen) in the various museums that have housed it over the years (Fig. 2.4).

By all accounts, Borissiak was an interesting character. A student of the pioneering Russian paleontologist Wladimir Kovalevsky (who first worked out the sequence of horse evolution in Europe in the 1870s), he received his training during the Czarist regime, but was about 45 years old when the Russian Revolution changed everything. Already a prominent scientist in the old royal academy and museum, he was the obvious choice to head the new Soviet Institute. He helped move the Institute from Leningrad to Moscow and later built it into one of the foremost paleontological collections in the world, the Paleontological Institute (known as the PIN for its Russian initials). Trained primarily as a zoologist, Borissiak also spent much of his career trying to liberate Soviet paleontology (in Borissiak's words, the "handmaiden of geology") from its long subservience to the needs of geologists and establish it as a distinct discipline with stronger ties to biology than geology.

By the time the Germans invaded the Soviet Union in 1941, Borissiak's Paleontological Insitute was well established and powerful within the Soviet bureaucratic structure, largely due to Borissiak's efforts. As Bodylevskaya (2008) describes it, the war years were terrible for the entire Soviet Union, especially when the Germans besieged Moscow. Much of the Institute was moved to Alma-

Figure 2.4. Two views of the most complete indricothere skeleton known. It is displayed in Moscow, based on specimens found north of the Aral Sea in the 1920s, plus a cast of the American Museum skull found in Mongolia (shown in Fig. 6.1). (Images courtesy M. Fortelius.)

Ata in Kazakhstan, while Borissiak was evacuated to the town of Frunze for his protection, though other scientists stayed behind in the museum to try to protect it and its contents. The Paleontological Institute survived the bombing and shelling and the other horrors of war and emerged stronger than ever. However, Borissiak himself died in 1944 at the age of 76, so he did not live to see the liberation of the Soviet Union or the fall of Nazi Germany in 1945. His role in the founding of Soviet paleontology was so prominent that today the organization he founded and nurtured through twenty-five years is named in his honor. Today, it is known as the A. A. Borissiak Paleontological Institute of the Russian Academy of Sciences.

Borissiak, Pavlova, and later Vera Gromova (1959) continued to collect and describe more indricothere material from the Aral Sea region and elsewhere in the former Soviet Union, along with more specimens from Mongolia. Today, their collection is one of the largest in existence and significant parts of it have still not been fully described or published. In recent years, the breakup of the Soviet Union has opened some of the former Soviet republics to exploration by Western scientists. Robert J. Emry of the Smithsonian Institution and Spencer G. Lucas of the New Mexico Museum of Natural History have conducted a number of field seasons in the former Soviet Republic of Kazakhstan, where they have accumulated many more specimens of indricotheres and other large mammals from the Eocene-Oligocene beds there (see Chapter 3). Thus, the effort in central Asia continues, although it is always subject to the vagaries of politics and funding.

Monsters of the Middle Kingdom

Although they were relatively late on the scene of vertebrate paleontology, China has also made important strides in the study of mammalian fossils, including indricotheres. During the early twentieth century, Chinese paleontology was largely under the supervision of foreign paleontologists. One of the most famous was the expedition by the American Museum's Walter Granger (discussed in Chapter 1) and Swedish geologist Johann Gunnar Andersson in 1921. They visited "Dragon Bone Hill," a cave deposit in Zhoukoudian outside Beijing, which they had originally discovered by tracing down the source of the "dragon bones" that were sold in Chinese apothecary shops (actually mammal fossil bones that were poached and smuggled and then ground into folk "medicines"). There they found a number of Ice Age mammal fossils, including a gigantic rhino-sized tapir, *Megatapirus*. Their work gave Chinese paleontology its first boost into the modern era. Under the supervision of Andersson's assistant, Otto Zdansky, Canadian anatomist Davidson Black, and French paleontologist and Jesuit priest Pierre Teilhard de Chardin, hundreds of Chinese laborers excavated the cave deposit in 1923 and the following years, and by 1926 they found the first remains of what became known as "Peking man" (now *Homo erectus*). By 1928 they had skulls, jaws, and

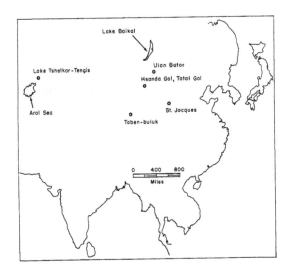

partial skeletons, and, using an $80,000 grant from the Rockefeller Foundation, they set up the Cenozoic Research Laboratory of the Geological Survey of China at Peking Union Medical College.

As the excavations at Zhoukoudian continued, Black worked on the "Peking man" fossils, but died suddenly in 1934 with the specimens still on his desk, largely undescribed. His job fell to a German Jewish anthropologist, Franz Weidenreich, who completed the descriptions. But in 1937, the Japanese invaded China and occupied Beijing, and the excavations ceased. For safekeeping, the specimens were locked up in safes in an American-run hospital, on the assumption that the Japanese would never interfere with an American facility. Japan was still officially neutral toward the U.S. (which was not involved in the Sino-Japanese war), and it would have caused a diplomatic crisis if they did. However, in the summer of 1941, Weidenreich could see the war clouds building between Japan and the U.S. and ordered casts made of the "Peking man" specimens, which were shipped to museums around the world as a precaution. To protect them from the Japanese, Weidenreich's secretary, Hu Chengzi, packed up the original fossils and hid them again, while Weidenreich was forced to leave China. In November 1941, anticipating the outbreak of war (the attack on Pearl Harbor occurred just a few weeks later on December 7), the specimens were secretly removed and transported by a contingent of U.S. Marines and Chinese scientists with the intent of smuggling them out of the Chinese port city of Qinhungdao. The fossils never arrived, and it is still a great mystery where they went. Fortunately, we still have the excellent casts, plus much more complete material excavated from Zhoukoudian in recent decades (starting in 1966), so the Chinese *Homo erectus* is one of the best documented of all human species.

As they were working on the Zhoukoudian human fossils in 1922, Teilhard de Chardin was sent by the French Museum of Natural History to the region of Ordos in Inner Mongolia to look for fossils. In the desert where the mighty

Figure 2.6. Zhou Ming-Zhen, also known as Minchen Chow. (Photo courtesy Wang Xiaoming.)

Huanghe (Yellow) River forms a big northern loop, he found an Oligocene locality near the French Catholic mission of St. Jacques (Fig. 2.5). There he collected numerous specimens, including indricotheres, carnivores, and rodents (Teilhard de Chardin, 1926). The localities were lost for some time until the Chinese rediscovered them in 1978 and 1979, collected new fossils, and built much larger and better collections there (Russell and Zhai, 1987, pp. 297–300). As noted by Teilhard de Chardin (1926), the fauna is very similar to that of the late Oligocene Hsanda Gol and Loh formations further north in Mongolia, described by Matthew and Granger (1923).

Among the talented Chinese paleontologists who worked alongside Black and Weidenreich was Yang Zhong-Jian (often known in English as C. C. Young), who led the Zhoukoudian excavations from 1928 to 1933. Yang and his co-workers continued their efforts to advance Chinese paleontology as best they could, although conditions were difficult during the Japanese occupation from 1937 until 1945. After the Japanese had left, China was consumed with the long-running struggle between Chiang Kai-Shek's Nationalists and Mao Zhe-Dong's Communists, which ended in the Communist victory in 1949. Nevertheless, Yang and his colleagues persevered, and they discovered some of the most important dinosaurs to come out of China, including the Triassic prosauropods from the Lufeng beds (*Lufengosaurus* and *Yunnanosaurus*), huge sauropods (*Mamenchisaurus*), and duckbilled dinosaurs (*Tsintaosaurus*) found largely in the 1940s and 1950s. Today, Yang is regarded as "the father of Chinese vertebrate paleontology" for his tireless efforts in keeping the profession going despite occupation, two wars, and political turmoil.

After Mao and the Communists had taken over, the Cenozoic Research Laboratory was affiliated with the National Planning and Steering Commission for Geological Works. By1953 it was an independent research unit, the Laboratory of Vertebrate Paleontology of Chinese Academy of Sciences. In 1957, its name was changed to the Institute of Vertebrate Paleontology, and in 1960 it became the Institute of Vertebrate Paleontology and Paleoanthropology (IVPP), which has been its identity for the past fifty-two years.

As the lab grew in size, it added paleontologists who studied a wider range of the Chinese fossil record. Among these was Zhou Ming-Zhen (often known in English as Minchen Chow), who became the driving force behind the study of fossil mammals in China in the later twentieth century (Fig. 2.6). According to the biography by Miao et al. (2010), Zhou was born in Shanghai on November 9, 1918, to a relatively well-educated family (his father was a math professor at a university in Shanghai). Even though he was extremely bright, he goofed off and got poor grades until his family threatened to disown him. After that, he became a stellar student and graduated in 1943 from Chongqing University, where he was a student of Yang Zhong-Jian, even as World War II was raging in Asia. He then took a job with the Sichuan Geological Survey and then at the Taiwan Geological Survey in 1946.

To get a better education, he traveled to the United States, where he earned his master's at Miami University in 1948 and his doctorate at Lehigh University in Pennsylvania in 1950. While he was there, he took summer classes at the American Museum, where he met the prominent vertebrate paleontologists of the time (including George Gaylord Simpson, Edwin Colbert, and Bobb Schaeffer) and worked in the Bighorn Basin Eocene and Paleocene beds with Glenn Lowell Jepsen of Princeton.

Now that he was hooked on fossil mammals and more highly trained in American fossil mammals, he returned to China in 1951. After taking a post at Shandong University and teaching there for a short time, he joined the IVPP in 1952, while it was still called the Cenozoic Research Laboratory. During the 1950s and 1960s, he led a number of important expeditions to Mongolia (in collaboration with the Soviets) as well as other parts of China and was the most important mentor to the next several generations of Chinese paleomammalogists. Before the Cultural Revolution, he had published over one hundred papers on fossil mammals.

More fieldwork by Zhou and other Chinese paleontologists in Inner Mongolia (the southern, or Chinese part of Mongolia, also known as Nei Monggol) yielded the first indricothere specimens found there since the Andrews-AMNH expeditions in the 1920s. In 1959, the Chinese and Soviet paleontologists were collecting in the middle Eocene Shara Murun beds of Mongolia, where they found an even smaller and more primitive indricothere, *Juxia sharamurunense,* described by Zhou and Chiu Chan-Siang in 1964. In 1963 Zhou and Chiu described the early Oligocene indricothere, *Urtinotherium incisivum,* which is slightly smaller than the giants of the later Oligocene. Other specimens of giant indricotheres from Mongolia and eventually other parts of China began to accumulate in museums. They were most recently reviewed in a large monograph by Qiu and Wang (2007).

Zhou and Zhang and their colleagues had to cope with enormous challenges. In 1966, after the failure of the "Great Leap Forward" to stimulate Chinese agricultural and industrial productivity, Mao launched a "Cultural Revolution." This movement sought to expunge all Western capitalist and bourgeois elements from Chinese society and make the nation a pure Chinese Communist utopia. In practical terms, it meant persecution of Mao's political opponents and nearly all of China's elites, especially those with a Western education. Most of China's educated population were imprisoned or sent to work in labor camps, and many were executed or forced to escape into exile. Historical relics and artifacts that evoked China's pre-Communist past were destroyed, and cultural and religious sites were ransacked for being insufficiently reflective of Marxist-Maoist ideology.

Naturally, the loss of all this talent constituted a "brain drain" on an enormous scale. It crippled the Chinese economy and society and made conditions even worse in China. The scientists at the IVPP feared for their lives, and some were dragged off to work in the rice paddies or to their deaths in prison. Most of the paleontologists survived by lying low and doing their work in secret, without any contact with the outside world. Thus, the Cultural Revolution transformed Chinese paleontology as it did the entire nation. The Chinese scientists I've spoken with are afraid to mention this dark time, or if they do, it is very brief and painful, just as shell-shocked veterans dread to tell the rest of the world their horror stories because they do not want to relive those memories. Zhou's biography (Miao et al., 2010) barely mentions the period between 1966 and 1976 at all.

Despite their troubles with the Cultural Revolution, the IVPP and its scientists managed to quietly keep working on materials that they had already collected and to make new collections, even though they were unable to read much about comparable fossils from the rest of the world or to travel outside China to see other scientists or to examine relevant fossils. Mao officially declared the Cultural Revolution over in 1969, but the political instability persisted until 1976, when the "Gang of the Four," which had actively pushed the "Cultural Revolution" during Mao's senility, was arrested and lost its power to terrorize Chinese society. The "Gang of the Four" consisted of Mao's last wife Jiang Qing and her associates Zhang Chunqiao, Yao Wenyuan, and Wang Hongwen, plus General Lin Biao. Their arrest shortly after the death of Mao in 1976 brought the reign of terror to an official end, and shortly thereafter the more moderate, West-friendly allies of Deng Xiaoping took power.

Slowly the scientists of the IVPP began to develop their contacts with Western scientists in the late 1970s and the 1980s. I vividly remember this period, because at each annual meeting of the Society of Vertebrate Paleontology we would be amazed by Chinese scientists presenting talks about stunningly complete and highly unusual specimens that had been unknown to the rest of the world for decades, even though the Chinese may have collected them long before the Cultural Revolution. Today the Chinese fossil beds are producing some of the most important discoveries, from bizarre fossil mammals to the amazing Liaoning fossils (famous for being articulated complete skeletons with feathers and stomach contents) that document the transition from birds to dinosaurs. There have been crucial discoveries, like the discovery at Doushanto of the earliest animal embryos and the discovery of a section that documents the details of the Earth's greatest mass extinction 250 Ma (million years ago).

Zhou Ming-Zhen, by now the senior member of the Chinese paleontology community that had survived the Cultural Revolution, served as Director of the IVPP from 1979 to 1984. He co-founded the journals *Vertebrata PalAsiatica* and *Palaeontologica Sinica* and edited both for many years. He made many trips abroad and worked hard to end the long isolation of Chinese paleontologists, bringing many prominent foreign paleontologists to visit and work in China. He was recognized for his role in promoting the survival and advancement of Chinese vertebrate paleontology with an honorary membership in the Society of Vertebrate Paleontology (SVP) in 1979 and with the Romer-Simpson Medal, the highest honor of the SVP, in 1993. Meanwhile, in China he was honored as an Academician of the Chinese Academy of Sciences in 1980.

Even though China had long been isolated, Zhou tried to quickly bring Chinese paleontology up to speed and introduced Western theoretical ideas, like cladistic systematics during the 1980s, while they were still controversial and debated within the Western scientific community. He urged the best and brightest young Chinese scientists to study abroad and pursue their advanced degrees at American universities, and today many top positions are held by his former students and their protégés (Luo Zhexi at the Carnegie Museum of Natural History in Pittsburgh, Wang Xiaoming at the Natural History Museum of Los Angeles County, Meng Jin at the American Museum of Natural History in New York). They, in turn, have brought Western scientists in as collaborators on Chinese explorations and research, further increasing the exchange between the two countries, although there are still significant hurdles, including the language barrier. It helps if Western paleontologists try to learn Mandarin, but they need to spend significant months or years in China to make real progress.

Zhou himself died in 1996 at the age of 78, having made a larger impact on Chinese mammalian paleontology than any other individual in Chinese history. The Qiu and Wang (2007) indricothere monograph was dedicated to him on the tenth anniversary of his passing. As Miao et al. (2010) put it, "he lived to see Chinese vertebrate paleontology thrive, and could have said, 'I came, I suffered, I survived and even succeeded.'"

The Giant Hunting Continues

This brief review of the major research efforts to find and study the fossils of indricotheres is by no means complete, nor is the story over. There are Chinese indricothere specialists still working in the IVPP, and new specimens continue to be discovered and described. The Soviets and Chinese collaborated in the 1950s with major expeditions to Mongolia, and the Chinese have worked hard in Nei Monggol ever since, while discovering many new localities throughout the rest of China. The Polish-Mongolian expeditions of the 1960s were famous for their dinosaur discoveries in Mongolia, but they found indricotheres too.

More recently, the Lucas and Emry expeditions to Kazakhstan reopened research areas that had been abandoned long before Kazakhstan broke away from the Soviet Union. In Chapter 3, we shall review the efforts of French paleontologists Pierre-Olivier Antoine, Jean-Loup Welcomme, and others to reopen the quarries in the Dera Bugti area and other parts of Pakistan, even though these tribal areas are extremely dangerous now because of warfare between tribal chiefs and the Pakistani government, Taliban insurgents, Islamic extremists, and the spillover of the military conflict in Afghanistan. Antoine and his colleagues have also extended their expeditions to Turkey and other areas in Asia, finding more and more new material of these extinct wonders. Those discoveries will be the subject of later chapters.

3

Lands of the Giants

Beasts of Baluchistan

The first indricothere fossils were found in the Baluchistan (also spelled Balochistan) region of what was then British India (Fig. 2.1) and now part of southern Pakistan. A soldier named Vickary brought back the first specimens in 1846, but they were so fragmentary that no one knew what they were. The geological and paleontological research there started with William Thomas Blanford in 1882–1883, Guy Pilgrim's work in 1907–1908 (Pilgrim, 1908, 1910), and Forster Cooper's expedition to Dera Bugti in 1909 and 1910 (Forster Cooper, 1911, 1913a, 1913b, 1923a, 1923b, 1934). Very little detailed geologic mapping of the region was done by these pioneers, since they were still conducting reconnaissance geological investigations without the time or resources to complete detailed modern maps or stratigraphic sections of the fossil-bearing beds. Blanford (1883) and Pilgrim (1910) published only minimal descriptions of the Bugti beds, and Forster Cooper (1911, 1913a, 1913b, 1923a, 1923b, 1934) made almost no mention of the geology at all in his many publications on the fossils found there. After Forster Cooper's last expedition to the region in 1911, almost no further collecting or research was done in the area until the late 1970s and 1980s, and the geological story of this region was still poorly known and confusingly interpreted. What ancient environments were represented in the Bugti beds where these creatures lived? What other fossil mammals were found with the indricotheres, and what do they tell us about the ancient environments? How old were the deposits? Some claimed they were early Miocene, while others placed them in the Oligocene.

Fortunately, new geological research and paleontological collecting at Dera Bugti has been conducted by French and local Baluchi workers, beginning with the efforts of Jean-Loup Welcomme and his colleagues in 1995 and 1996. This has resulted in a large volume of research as well as new specimens, better stratigraphic control of the fossils, and a much clearer idea of their age. They are sum-

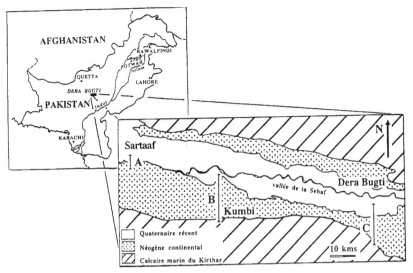

Figure 3.1. Index map of the Dera Bugti localities in the Baluchistan region of Pakistan.
(Courtesy P.-O. Antoine).

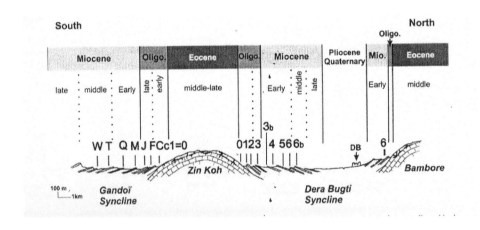

Figure 3.2. Structural cross-section showing the anticlines and synclines of the Dera Bugti region. (Courtesy P.-O. Antoine.)

marized by Welcomme et al (1997, 2001), Welcomme and Ginsburg (1997), and
Métais et al. (2009).

The current understanding of the section is shown in Figures 3.1 and 3.2.
The Dera Bugti beds occur in a large folded sequence of rocks, with anticlinal
ridges (Zin Koh Range and Bambore Range) that exposed middle-upper Eocene
beds and synclines (Dera Bugti, Gandoï), which are filled with Oligocene,
Miocene, Pliocene, and Quaternary deposits (Fig. 3.2). At the base of the sequence is the marine Eocene Kirthar Formation, first described by Blanford in
1879 (Fig. 3.3). It contains marine vertebrates (mainly shark teeth), a rich mol-

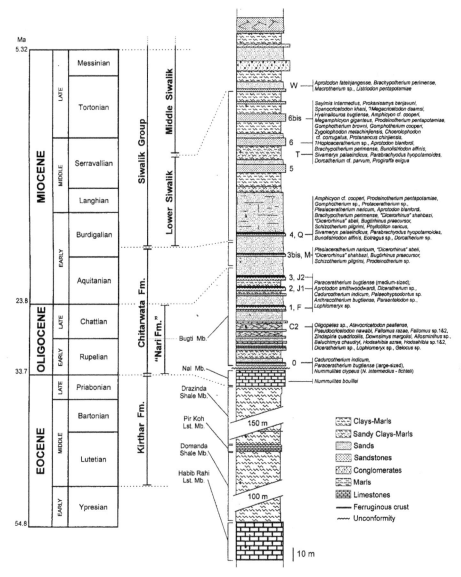

Figure 3.3. Lithostratigraphy and biostratigraphy of units within the Dera Bugti area. (Courtesy P.-O. Antoine.)

luscan fauna (especially scallops and oysters), and the coin-sized foraminiferans known as nummulitids, whose rapid evolution and great abundance all across Asia, northern Africa, and Europe make them the preferred index fossils for the Eocene. The Kirthar Formation is divided into several distinct members that appear to span the middle and late Eocene.

Overlying the Kirthar Formation is the Chitarwata Formation, which is approximately 70 m thick and appears to span the early Oligocene to the early Miocene in age. The lower or basal part of the Chitarwata Formation is called the "Nari Formation" in some references. According to Welcomme et al. (2001),

Figure 3.4. Outcrops of Chitarwata Formation in the Dera Bugti area. (Courtesy P.-O. Antoine.)

the Nal Member (or basal Nari Formation) grades conformably into the marine Kirthar Formation beneath it, establishing an almost complete Eocene-Oligocene transitional sequence. The lower Nari is a massive white to gray crystalline limestone with a mixture of marine fossils (the early Oligocene foraminifers *Nummulites bouillei* and *N. clypeus,* plus scallops, oysters, sharks, and sea cows), and rare terrestrial mammals, including one of the last of the amynodont rhinos *Cadurcotherium indicum* and large specimens of *Paraceratherium bugtiense* (Antoine et al., 2004). The discovery of this mixture of early Oligocene marine fossils and terrestrial mammals answers the longstanding controversy over the age of the lower Chitarwata Formation. It also clarifies the confusion of many authors (e.g., Prothero and Schoch, 1989; Raza and Meyer, 1984; Lucas et al., 1998), who were puzzled at how indricothere and *Cadurcotherium* specimens could occur so late at the early Miocene (possibly as young as 15 Ma), when most amynodonts lived in the middle and late Eocene and vanished in North America and China after the early Oligocene. Thanks to this new stratigraphic data, it is now clear that cadurcotheres and indricotheres didn't survive to the early Miocene but were restricted to the Oligocene (as the late amynodonts are in other continents).

The remaining parts of the Chitarwata Formation (Figs. 3.3, 3.4) appear to be middle Oligocene through early Miocene in age. According to Métais et al. (2009), the bulk of the Chitarwata Formation consists of coastal deltaic deposits at the base, which grade into river channel and floodplain environments near the top of the section. This sequence was produced by deltas building out from the early uplift of the Himalayas, which rose even more rapidly in the Miocene to form the foreland basin sequence of the Siwaliks. According to Martin et al. (2011), the climate supported a dense but dry subtropical to temperate forest. The Bugti Member (or Upper "Nari Formation") produces a range of vertebrates, from fish and turtles and crocodiles to a rich Oligocene fauna of rodents, a few of which are closely related to those found in the Hsanda Gol Formation in Mongolia. The large mammals include not only *Paraceratherium bugtiense* (Fig. 3.5), but the youngest known specimens of the last of the amynodonts, *Cadurcotherium,* true rhinocerotids (*Aprotodon, Epiaceratherium, Diaceratherium*), a small elephantoid proboscidean, huge pig-like entelodonts (*Parentelodon*) (Fig. 3.6A), hippolike anthracotheres (*Anthracotherium bugtiense*) (Fig. 3.6B), early pigs (*Sanitherium, Pecarichoerus, Hyotherium,* as described by Orliac et al., 2010), primitive ruminant artiodactyls (*Palaeohypsodontus, Lophiomeryx, Gelocus, Microbunodon, Iberomeryx, Dremotherium*) (Fig. 3.6C), an opossum (*Asiadidelphis*), and even lemur-like, tarsier-like and monkey-like primates (*Bugtilemur, Bugtipithecus, Guangxilemur, Phileosimias,* according to Marivaux et al., 2005).

These fossil faunas are surprisingly unbalanced and depauperate, with only a handful of predatory mammals that are still under study by Stéphane Peigné (P.-O. Antoine, pers. comm., 2011). The provisional list includes a *Stenoplesictis*-like feloid (cat relative), four basal marten-sized arctoids (bear relatives), and a couple of hyaenodont creodonts (Fig.3.6C). The collections consist mostly of big rhinos and indricotheres and larger artiodactyls, and tiny rodents, primates, and opossums. Since a mammalian fauna with few predators is very unlikely on ecological grounds, this probably represents the limited collecting and sampling that has been done up to this point (as already noted by Marivaux et al., 2005, who pointed out that the fauna was dominated by very large and tiny mammals, with a lack of medium-sized mammals so far).

The uppermost part of the Chitarwata Formation yields an early Miocene mammalian fauna (Fig. 3.3). This sequence is not very thoroughly sampled yet, but it already produces characteristically early Miocene taxa such as the rhinocerotids *Plesiaceratherium naricum, Pleuroceros blanfordi, Mesaceratherium welcommi, "Dicerorhinus" abeli, Prosantorhinus shahbazi,* and *Bugtirhinus praecursor,* as well as the chalicothere *Schizotherium,* and the huge proboscidean with downturned tusks, *Prodeinotherium.* Immediately above these

Figure 3.5. Piles of broken indricothere bones eroding from rocks in the Dera Bugti area. (Courtesy P.-O. Antoine.)

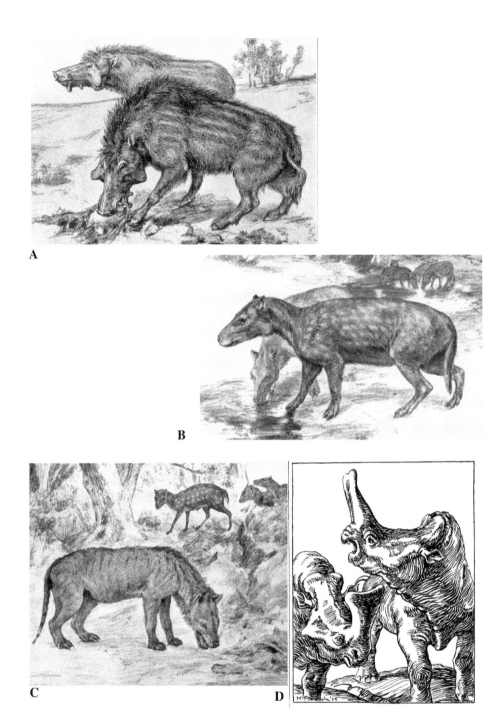

Figure 3.6. Reconstructions of typical mammals of the late Eocene and Oligocene in Asia and North America. A. The huge pig-like creatures known as entelodonts; giant Oligocene entelodonts like Asian Paraentelodon and American Daeodon (or Dinohyus) were the size of living rhinos. B. The pig-like anthracotheres, an extinct group thought to be related to hippos and whales. C. The creodont predator Hyaenodon with deer-like ruminants in the background. D. The giant battering-ram horned brontothere Embolotherium. (A–C by R. Bruce Horsfall, in Scott, 1930; D after Osborn, 1929.)

lower Miocene deposits of the upper Chitarwata are the first good Siwalik faunas, which have an excellent record of both small and large mammals (Fig. 3.3).

The recent French-Baluchi work in the Dera Bugti region has greatly clarified our understanding of the geology and depositional environments, and also appears to resolve the confusion about the age of the fossils. Crucial new specimens, such as the primates reported by Marivaux et al. (2005), have greatly increased the interest in these deposits and their fossils. However, the collections are still too small and seem to be lacking any medium-sized mammals and sufficient numbers of predators that we would expect from the Oligocene in Asia. Some of this may be a preservational bias, since the sediments are more coarse-grained and deposited in high-energy rivers and streams, which tend to break up the smaller and more fragile fossils. Because the political situation in Pakistan is now very precarious, it will be a long time before scientists are allowed into Dera Bugti again.

Monsters of Mongolia

Although the first specimens of indricotheres were reported from Baluchistan, the most complete specimens and most heavily publicized discoveries were made in Mongolia starting with the 1922–1930 American Museum expeditions discussed in Chapter 1. Berkey and Morris (1927) were the geologists on those expeditions and laid down the original geologic framework on which later work was built. They recognized a distinct sequence of formations in Mongolia, starting with the upper Paleocene Gashato Formation, which produces a number of fascinating mammals that demonstrate the earliest radiation of groups like rodents and tethytheres in the late Paleocene. Overlying it is the Naran Buluk Formation, which yields a variety of early Eocene and middle Eocene mammals.

The next unit was the Ergilin-Dzo Formation (called the Elegan Formation by Höck et al., 1999 and Daxner-Höck and Badamgarav, 2007), which had long been considered early Oligocene. It is chock full of amynodonts and especially the huge battering-ram brontotheres known as embolotheres (Fig. 3.6D). It is about 50–80 m thick, composed mainly of light-colored quartz sands and mudstones, with intercalations of red sand and clay at the top. This formation and its mammal faunas became the basis for the Ergilian land mammal age in Asia, and it also yields indricotheres such as *Urtinotherium* and the earliest "*Indricotherium*" (or *Paraceratherium*).

Traditionally, this unit had been considered early Oligocene, as had its correlatives units and faunas in China. However, when Carl Swisher produced new argon-40/argon-39 (^{40}Ar/^{39}Ar) dates on the "early Oligocene" Chadronian land mammal "age" in North America, and these dates were combined with my magnetic stratigraphy of the key sequences, the Chadronian proved to be late Eocene in age. As a result, the entire time scale in North America needed to be re-adjusted (Swisher and Prothero, 1990; Prothero and Swisher, 1992; Prothero and Berggren,

1992). The "late Eocene" Uintan and Duchesnean land mammal ages became middle Eocene, the "early Oligocene" Chadronian became late Eocene, and the "middle-late Oligocene" Orellan and Whitneyan became early Oligocene.

Although it took a while for American paleontologists to unlearn their early training that "Chadronian equals early Oligocene," and for Asian paleontologists to do the same (e.g., Wang, 1992, and Dashzeveg, 1993, still used the old correlations), over the past twenty years both American and Asian paleontologists have come to regard the Chadronian in North America (Woodburne, 2004; Prothero and Emry, 2004) and the Ergilian land mammal age in Asia (Ducrocq, 1993; Tong et al., 1995; Meng and McKenna, 1998) as late Eocene.

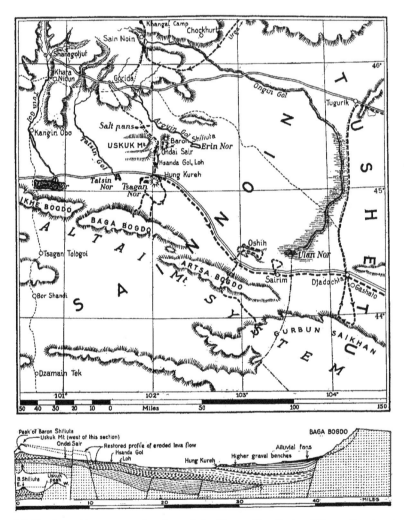

Figure 3.7. Stratigraphic cross-section through the Tsagan Nor Basin of Mongolia, showing the Hsanda Gol and Loh formations. (After Berkey and Morris field notes, published by Osborn, 1929.)

One of the most striking results of this change in the time scales is that all the huge brontotheres of both North America and Asia, the largest land mammals that had ever lived before the Oligocene, have been re-correlated with the late Eocene, so there are no longer any Oligocene brontotheres (including the Asian embolotheres). This means that trade books and captions that show brontotheres almost always state that they come from the Oligocene, but this is no longer true. Most of these popular books and websites and toys for the mass market also use outdated names for these brontotheres, such as *"Brontotherium," "Brontops," "Titanotherium," "Menodus,"* and other names that were propagated by the hypersplitting in Osborn's (1929) titanothere monograph. None of these names are valid, according to Mihlbachler's (2008) revision of the group, and only *Megacerops coloradense* remains as a valid name for Chadronian brontotheres (Mihlbachler et al., 2004).

Overlying the Ergilin-Dzo in many parts of Mongolia are the Houldjin gravels, a local unit that is often considered a basal member of the overlying Hsanda Gol Formation. Höck et al. (1999), Bryant and McKenna (1995), Daxner-Höck and Badamgarav (2007), and Kraatz and Geisler (2010) did not even mention the Houldjin gravels in their more recent stratigraphic summaries, since these coarse-grained deposits do not appear at the base of the Hsanda Gol in every part of Mongolia. Because of the high energy of these coarse-grained flood channel deposits, most of the fossils are fragmentary and badly damaged from being rolled around in gravels. Matthew and Granger (1923) reported the amynodont *Cadurcotherium,* as well as rhinocerotid and indricothere teeth, huge entelodonts, and tortoise shell fragments. However, recent work by Banyue Wang and colleagues has greatly improved on the short faunal list reported by Matthew and Granger (1923). Wang (2007) and Wang et al. (2009) reported a much more diverse fauna, including lagomorphs, rodents, hedgehogs, true rhinocerotids, the amynodont *Cadurcodon,* indricotheres (called *Aralotherium* by Wang et al., 2009), the creodont *Hyaenodon,* huge entelodonts, and abundant brontotheres. Wang et al. (2009) assign the Houldjin gravel fauna to the late Eocene (Ergilian), since it not only shares many taxa in common with other late Eocene Mongolian faunas like Ergilin Dzo and Ulan Gochu, but also contains brontotheres, which died out at the end of the late Eocene.

The main Oligocene unit throughout Mongolia is the Hsanda Gol Formation (also spelled "Shandgol"). As currently recognized by Höck et al. (1999), Bryant and McKenna (1995), Daxner-Höck and Badamgarav (2007), and Kraatz and Geisler (2010), the Hsanda Gol Formation consists of about 80 m of redbeds, floodplain mudstones, and gray fluvial sands (Figs. 3.7, 3.8) that intertongue with the underlying Ergilin-Dzo where the Houldjin gravels are not present. There are also abundant dune deposits whose sand content was reworked from the underlying units (as is true of the Oligocene in North America as well). Höck et al. (1999) and Kraatz and Geisler (2010) divided the Hsanda Gol Formation into a lower Tatal Member and an upper Shand Member, which are separated by the

A

B

C

*Figure 3.8.A. View of the Hsanda Gol Formation from the Bryant's Hill locality. Dark rocks in center are the dated lava flows. B. Southern ridge at Loh, showing the Loh Formation. Foothills of the Gobi Altai Mountains in the background. C. Close-up of the section at Loh, with a fossil rodent skull (*Tsaganomysaltaicus*) in the foreground. (Photos courtesy B. Kraatz.)*

marker basalt lava flow layer dated at 31.5 Ma. The lower portion of the sequence (from the base to the lower Shand beds just above the basalt) produces the early Oligocene Ulaan Khongli fauna. The uppermost part of the Shand Member yields the later Oligocene Zavila fauna. Daxner-Höck and Badamgarav (2007) recognized two rodent "biozones" in the Hsanda Gol Formation that they correlated with the early and late Oligocene.

The Hsanda Gol Formation was traditionally considered upper Oligocene, but with the Ergilian moved from the early Oligocene to the late Eocene, the Hsanda Gol Formation includes lower Oligocene beds as well (Fig. 3.9). This was recently established by Daxner-Höck and Badamgarav (2007), who correlated the rodents to early Oligocene biozones Mammal Paleogene (MP) 21–24.

Further corroboration came from work by Kraatz and Geisler (2010), who used magnetic stratigraphy combined with new dates on the lava flow in the middle of the section. This technique, developed in the 1970s, allows the geologist to date rocks by the pattern of the flip-flops of the earth's magnetic field as recorded in the sedimentary sequence. Kraatz and Geisler (2010) found that the Hsanda Gol Formation ranged in age from the end of magnetic Chron C13n (early Oligocene, 33.4 Ma in the timescale of Gradstein et al., 2004), spanned most of the long reversed magnetic Chron C12n (31.2–33.4 Ma), and terminated in the late Oligocene Chron C12n (31.0 Ma), based on the 31.5 Ma date of the reversed polarity basalt flow in the middle of the sequence (Fig. 3.10).

Overlying the Hsanda Gol Formation is the Loh Formation, composed of about 30–50 m of pinkish and greenish clays and tan sandstones. Although the upper part of the Loh Formation produces a classic early Miocene mammal fauna, at least some of the lower part of the Loh Formation (and the correlative Taben Buluk locality in Mongolia) is late Oligocene in age (Meng and McKenna, 1998; Daxner-Höck and Badamgarav, 2007; Kraatz and Geisler, 2010) and yields late Oligocene taxa such as giant indricotheres, rhinocerotids, chalicotheres, plus characteristic rodents and lagomorphs (Fig. 3.9). Daxner-Höck and Badamgarav (2007) recognized five separate rodent biozones in the Loh Formation, ranging from the late Oligocene level MP25–30 to early Miocene biozones Mammal Neogene (MN) 3–5. Thus, the Tabenbulukian Asian land mammal age begins in the late Oligocene (Chron C12n, 31.0 Ma) and continues up into the early Miocene.

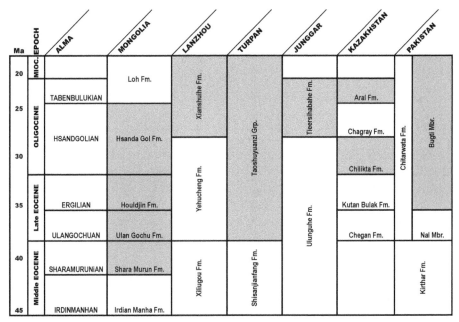

Figure 3.9. Correlation chart of late Eocene-Oligocene rocks and faunas of Asia, from Mongolia to China (Lanzhou, Turpan, Junggar) to Kazakhstan and Pakistan. Shaded boxes indicate indricothere-bearing deposits. (Modified by T.D. LeVelle from Qiu and Wang, 2007.)

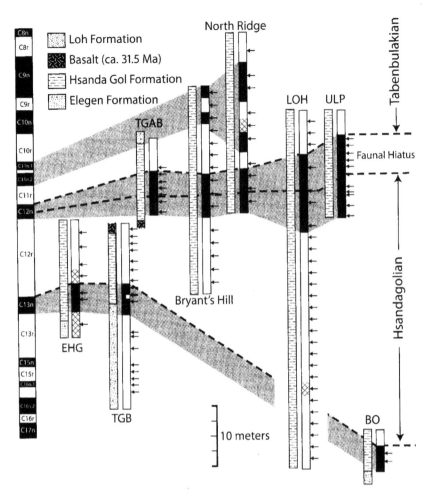

Figure 3.10. Magnetic stratigraphy of the Hsanda Gol and Loh beds. (After Kraatz and Geisler, 2010; courtesy B. Kraatz.)

The recalibration and redefinition of the classic sequence in Mongolia (Fig. 3.9) has led to rearrangement of the sequences of faunas as recognized in China by Russell and Zhai (1987), Wang (1992), and Dashzeveg (1993). Most of the Chinese localities, which yield no radiometric dates and are not amenable to magnetic stratigraphy, are correlated with the global time scale based on the Mongolian reference standard. These include a large number of different fossil quarries in many different Chinese provinces (Russell and Zhai, 1987; Wang, 1992). These localities have been recalibrated into the modern correlations by Tong et al. (1995) and Meng and McKenna (1998), so that the confusion of Eocene and Oligocene localities has now been corrected. However, one must be careful of reading and interpreting older literature based on these old correlations.

The re-correlation has also had the positive result that the Asian rocks and their faunas match the pattern seen in North America and in Europe across the

Eocene-Oligocene boundary (Meng and McKenna, 1998). This is crucial, since a miscorrelation and misunderstanding of the Eocene-Oligocene transition has long hampered research (Ducrocq, 1993) in what was going on during this crucial episode of earth history (see Chapter 7).

Mysteries of Kazakhstan

Until recently, very little about the geology of Kazakhstan was published in English, and it did not reflect modern concepts of stratigraphy as recognized by international codes of stratigraphic nomenclature. Yet there are fossils from all over Kazakhstan, and for the past few decades since Kazakhstan became independent with the end of the Soviet Union in 1990, a large number of Kazakh and American researchers have renewed research in the area. Much of the newly published research on the fossils and the rocks of Kazakhstan has been conducted by Spencer Lucas of the New Mexico Museum of Natural History, Robert J. Emry of the Smithsonian Institution, and their Kazakh colleagues (Emry et al., 1998; Lucas et al., 1998; Lucas and Emry, 1996a, 1996b, 1996c, 1999; Lucas and Bayshashov, 1996; Lucas and Bendukidze, 1997; Lucas et al., 1997). These new collections and new research have cleared up a lot of the confusion caused by early twentieth-century Soviet work, which did not use modern internationally accepted concepts of geology or systematics.

Emry et al. (1998) and Lucas et al. (1998) summarized the current understanding of stratigraphy of this region. They have modified the original stratigraphy of Borisov (1963), which recognized a series of "svitas" (the Russian term for a rock unit, roughly comparable to a formation in international codes of stratigraphy). Emry et al. (1998) provided a correlation of these svitas from the Zaysan Basin of western Kazakhstan (Fig. 3.9) with the standardized Asian chronology of east Asia (China and Mongolia), and from there these units can then be correlated with global time scales. Lucas et al. (1998) correlated the rocks and faunas from the Aral Sea region of western Kazakhstan,

The sequence in the Zaysan Basin of eastern Kazakhstan includes a number of middle Eocene svitas (Fig. 3.11) that do not contain indricotheres but are correlated with middle Eocene Irdinmanhan and Sharamurunian Asian land mammal ages of China and Mongolia. There is a thick Eocene-Oligocene sequence of units. Borisov (1963) called these (in order from bottom to top): the Kyzylkain svita, Aksyir svita, Kusto svita, and Buran svita. Emry et al. (1998) provide updated descriptions of each of these units, and point out that "svitas" are not true equivalents of formations (or mappable units defined by their lithology) as recognized by international stratigraphic codes. For that reason, they do not recognize separate Kusto and Buran svitas, but merge them into a single mappable Kusto-Buran svita.

Figure 3.11. Aerial photo of the exposures of Eocene-Oligocene beds north of the Aral Sea. (Courtesy S. Lucas.)

The Aksyir svita produces abundant fossils, especially of aquatic fish, turtles, and crocodilians, along with numerous specimens of small mammals, which are closely comparable to those from the Ergilian, or late Eocene, of China. The upper part of the unit produces sparse specimens of large mammals, including the hippo-like anthracotheres, the primitive ruminant *Archaeomeryx*, and the type specimen of the huge amynodont rhino *Zaisanamynodon borisovi*. Based on correlation with the Ulan Gochu and other Mongolian-Chinese localities, it appears that the fauna from the Aksyir svita is latest Eocene, or Ergilian, in age.

The Kusto-Buran svita produces a much richer mammal fauna, including numerous large mammals. These include the amynodont rhinos *Cadurcodon ardynense,* which ranges from the Ergilian to the Shandgolian in Mongolia, brontotheres (restricted to the late Eocene Ergilian and earlier formations in Mongolia and China), the pig-like *Entelodon* (which last appears in the Ergilian), the ruminant *Gobiomeryx* (restricted to the Ergilian), anthracotheres (closest to *Aepinacodon* of the late Eocene of North America), as well as hyaenodont creodonts. The small mammals (primarily rodents), on the other hand, suggest an early Oligocene (Shandgolian) correlation for the Kusto-Buran svita. Emry et al. (1998) conclude that the Ergilian-Shandgolian transition occurs within the upper part of the Kusto-Buran svita, and by extension, so does the Eocene-Oligocene boundary. This falsifies earlier notions of Soviet geologist that the entire Zaysan sequence was Oligocene. Instead, only the uppermost part of the upper svita is Oligocene, and the rest is upper Eocene (Ergilian).

In the Aktau Mountains just south of the Zaysan Basin in eastern Kazakhstan (Fig. 3.12) is another upper Oligocene sequence of rocks that produced the jaw of a huge indricothere (Lucas and Bayshashov, 1996; Lucas et al., 1997). Named *Paraceratherium zhajremensis* by Bayshashov, it is larger than most other species of *Paraceratherium,* and its jaw is deeper below the last lower molar than any other species, so it is considered a valid separate species. It comes from the Aktau Formation, an upper Oligocene unit that underlies an extensive sequence of lower Miocene mammal-bearing deposits.

Lucas et al. (1998) studied the thick sequence of rocks north of the Aral Sea (Figs. 3.11, 3.12). The stratigraphic nomenclature in this part of Kazakhstan differs from that in the Zaysan Basin to the east (Fig. 3.12). From bottom to top, the units are: the marine yellow-green bentonitic shale of the Chegan Formation, which contains late Eocene-early Oligocene dinoflagellate planktonic microfossils, turritellid snails, marine bivalves, and nautiloids, but no vertebrate fossils; the yellow sandstones of the Kutanbulak Formation (Fig. 3.11), which contains only plant fossils and no age-diagnostic invertebrates or vertebrate fossils; the pinkish shales and gray sandstones of the Chilikta Formation (Fig. 3.11), which yields fossils of *Paraceratherium* (Lucas and Emry, 1996a) as well as plant fossils including leaves and charophytes; the yellow-gray sandstones of the Chagray Formation, which produces a few mollusks and plant fossils; and the youngest unit, the green claystones and shales of the Aral Formation, which contains abundant corbulid bivalves, fish fragments, and a few cetacean fossils (including a delphinid dolphin). There is a large mammal fauna from the Aral Formation, including many diverse rodents and lagomorphs, hedgehogs, moles, shrews, plus a new genus of weasel. The large mammals include not only Borissiak's type spec-

Figure 3.12. Index map showing the location of major fossil mammal localities in Kazakhstan. (Courtesy S. Lucas.)

Figure 3.13. Distal fragment of a huge humerus referred to Paraceratherium, found in 2006 in middle Oligocene deposits of the Kizilirmak Formation near Gözükizilli, central Anatolia, Turkey. French-Turkish paleontologist Sevket Sen is at left and his Turkish colleague Ebru Albayrak at right. (Courtesy of P.-O. Antoine.)

imens of *Indricotherium transouralicum* (as well as specimens called "*Paraceratherium prohorovi*"), but also true rhinocerotids (*Aprotodon, Aceratherium, Protaceratherium,* and *Eggysodon*) and ruminants (*Lophiomeryx, Prodremotherium, Miomeryx, Amphitragulus,* and *Micromeryx*). Taken together, both the large and small mammals indicate a late Oligocene (Tabenbulukian) age of the Aral Formation fauna, as calibrated from the Mongolian-Chinese biochronology.

Thus, the questions about the age of the original material of *Indricotherium* described by Borissak are now answered, and localities between the Zaysan Basin of eastern Kazakstan and the Aral Sea region of the west provide an important Eocene-Oligocene land mammal sequence to supplement the records of other parts of Asia.

Talking Turkey

So far, the indricothere record was known only for central and eastern Asia, from Kazakhstan and Baluchistan to Mongolia and China. Nearly every region with good exposures of Oligocene terrestrial beds in central and eastern Asia seems to produce them. This all changed in 2002, when a group led by Levent Karadenizli, Gerçek Saraç, Sevket Sen of Turkey, and Pierre-Olivier Antoine of France found the first remains of indricotheres in Turkey, described by Antoine et al. (2008). The localities are in the upper Oligocene Kizilirmak Formation near Gözükızıllı, just southeast of Ankara in the Çankırı-Çorum sedimentary basin in central Turkey. The indricothere specimen is only a broken and damaged humerus (Fig. 3.13), but its large size and shape are unmistakable. Found with this specimen are fossils of the late Oligocene rhinocerotid *Protaceratherium albigense* and distinctive late Oligocene rodents (*Eucricetodon, Sayimys, Glirulus, Bransatoglis,* plus unidentified ctenodactylids and dipodids) of European and Asian affinities. Nearby localities produce such distinctive artiodactyls as *Palaeo-hypsodontus, Iberomeryx,* and *Dremotherium,* and other tragulids and cervoids, as well as carnivores, rodents, and lagomorphs.

Later on, Sen et al. (2011) described other gigantic postcranial remains referred to *Paraceratherium* from the Kagizman-Tuzluca basin in eastern Turkey, right next to the Armenian border. So far, giant rhinoceroses are reported from the upper Oligocene Güngörmez-Kizilkaya Formation, and crocodile teeth and small mammals were reported from the Miocene beds just above this unit.

These discoveries provide not only the furthest extension of the geographic range of indricotheres to the southwest, but also corroborate earlier reports of indricotheres from eastern Europe to the west of Turkey. Previously, there were fragmentary specimens reported from the former Yugoslavia, Bulgaria, and Romania (Nikolov and Heissig, 1985; Lucas and Sobus, 1989; Spassov, 1989; Codrea, 2000). In addition, Gabunia (1955, 1964, 1966) reported indricotheres from the Caucasus region of formerly Soviet Georgia, including the enigmatic *Benaratherium,* which most paleontologists regard as yet another junior synonym for *Paraceratherium*.

Thus, we now have indricotheres from the Oligocene in nearly every part of Asia, from China and Mongolia in the east, India and Pakistan in the south, Kazakhstan and Georgia in the former Soviet Union, and Turkey, Rumania, Bulgaria and parts of the former Yugoslavia in westernmost Asia and eastern Europe. However, as many authors have noted (e.g., Heissig, 1979), there is a sharp break between these Asian and eastern European faunas and those of western Europe at the same time in the Oligocene. Those differences will be explored in greater detail in Chapter 7.

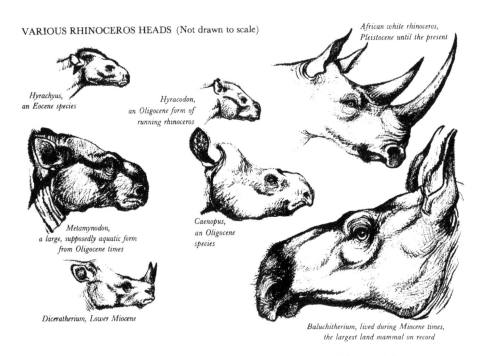

VARIOUS RHINOCEROS HEADS (Not drawn to scale)

Hyrachyus,
an Eocene species

Hyracodon,
an Oligocene form of
running rhinoceros

African white rhinoceros,
Pleistocene until the present

Metamynodon,
a large, supposedly aquatic form
from Oligocene times

Caenopus,
an Oligocene
species

Diceratherium, Lower Miocene

Baluchitherium, lived during Miocene times,
the largest land mammal on record

Figure 4.1. The diversity of head and horn shapes of different kinds of rhinos, all drawn to a common scale. (From Scheele, 1955.)

4

Rhino Roots

Rhinos without Horns

Before we look at the gigantic indricotheres in greater detail, we need to place them in the context of the evolution of the various types of rhinos (both extinct and surviving). Where did indricotheres come from? What were their closest relatives? What features allows us call these huge creatures without horns rhinoceroses?

The last question is the first misconception that we need to clarify. Horns occur in all five living species of fossil rhino, but they are only rarely found in a few lineages of fossil rhinos. This comes as a shock to most people who think "horn equals rhino" when they see one in the zoo or on TV and never notice the many other features that make rhinos distinct. What is up with that horn, anyway?

A rhino horn is not like that of a cow or sheep or antelope. Those creatures (the ruminants) have a horn made of a solid bony core surrounded by a sheath made from the same protein (keratin) found in your fingernails and hair. Nor are rhino horns like the ossicones of giraffes (which are solid bone with only a fleshy covering), nor like the antlers of deer (which are solid bone, but are grown and shed each year). A rhinoceros "horn" is actually made of dense fibers of hair glued together—there is no true bone within it at all. Thus, it grows throughout the rhino's life as does your hair or fingernails and breaks and wears down and abrades quite easily. Because it has no bone inside it, only perishable keratin, we almost never find the horn preserved on fossil rhinos. (The exceptions are the few examples of mummified woolly rhinos, which are preserved not only with complete horns, but even stomach contents, skin, and fur.) Instead, we must infer the size and shape of the horn from the roughened area on the top of the skull (nose or forehead or both). This indicates the point where the hairs of the horn glued in to the skull. We can see this pattern quite clearly on both living rhinos and also the extinct rhinos that had horns. Because the size of roughened area in-

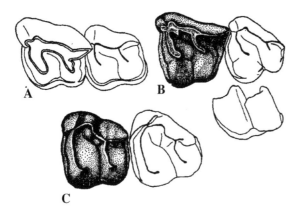

Figure 4.2. Occlusal view of the second and third left upper molars in different rhinocerotoid taxa (side view of lower molars below each pair). A. Amynodon, *a hippo-like rhinocerotoid. B.* Hyracodon, *a running rhinocerotoid. C.* Hyrachyus, *the most primitive known rhinocerotoid. Note how the second molars are shaped like the Greek letter π, and the third molars (on the right of figure 4.2B-C) lose the back outer crest (the metastyle), resulting in crests with a more V-shaped pattern. (From Radinsky, 1966.)*

dicates where a horn was once present, we are pretty sure that most extinct rhinos had no horn whatsoever. Since the horn is not a crucial feature in recognizing a rhino, we must look at other parts of the anatomy.

Rhinos diverged into a wide variety of sizes and body shapes as they adapted to a wide range of lifestyles (Fig. 4.1). Some were the size of a small dog, while the indricotheres were larger than most elephants. Some had typical rhino proportions, while others had long legs for running, and still others had short legs and squat hippo-like bodies for living in the water. Most rhinos had no horns, but some had a single horn on the nose (the living Indian rhino and many extinct forms), or a huge horn only on the forehead (a group called elasmotheres). Some had a pair of horns on the nose side-by-side (evolved independently in two different lineages), while others had horns in tandem, one behind the other on the nose and forehead (the living African rhinos). Some had a trunk or proboscis like a tapir or mastodont (with or without horns). In many parts of North American and Eurasia, rhinos are among the most common mammal fossils. Except when there were mastodonts or mammoths around, they were among the largest mammals on the landscape.

So if neither horns nor any rhino body shape are diagnostic of a fossil rhino, how can we tell if it's a rhino in the first place? There are many distinctive features of the skull and skeleton that allow a paleontologist to recognize a rhino, but the easiest and most distinctive features to recognize are its teeth. More than any other anatomical structure, mammalian paleontologists study and use teeth to identify mammal fossils. This is because teeth are covered with enamel (the hardest substance in your body), and are much more durable and likely to be preserved than any other part of the skeleton. In addition, teeth have a distinctive pattern of crests and cusps in nearly every group of mammals, reflecting their ancestry from a particular group with a unique tooth pattern. Superimposed on this basic inherited tooth pattern of their ancestors is the influence of what they ate as well. The teeth of carnivorous mammals are usually sharp blades for slicing meat or stabbing, pointed teeth for grabbing prey, while those of the herbivores have different distinctive patterns of crests for shredding tough vegetation. The teeth of omnivores have the primitive pattern of simple rounded cusps on the corners of the crown of the cheek teeth for processing a wide variety of food, from meat to vegetation.

In the case of rhinos, they adopted a cheek-tooth pattern that became stereotyped very early in their evolution about 50 Ma. Most rhinos have upper molars (the last three cheek teeth that erupt without replacing a "baby tooth") with three cross-crests forming a Greek letter pi (π) (Fig. 4.2). In addition, most advanced rhinos have premolars (the first three or four cheek teeth, which replace the baby teeth when the animal grows up) that also have a pi (π) pattern, or something that approaches it. By contrast, the lower molars have crown pattern that looks like a set of the letter L attached to one another (Fig. 4.3). There are details of the cross-crests, as well as the presence or absence of additional crests or cusps, the shape and angle of the crest, narrow shelf-like structures ("cingula") around the base of the tooth, and so on that help a paleontologist recognize specific rhinos, but the general pattern is pretty consistent within the entire group.

There are other details of the skull region (especially the base of the skull and

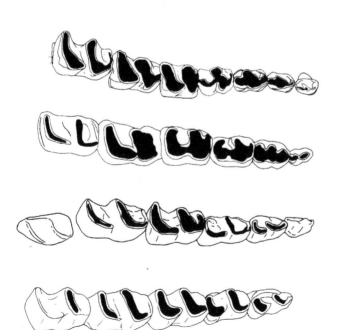

Figure 4.3. Rhino lower molars in crown view, showing the typical "L"-shaped crest pattern. (From Lucas and Sobus, 1989; courtesy S. Lucas.)

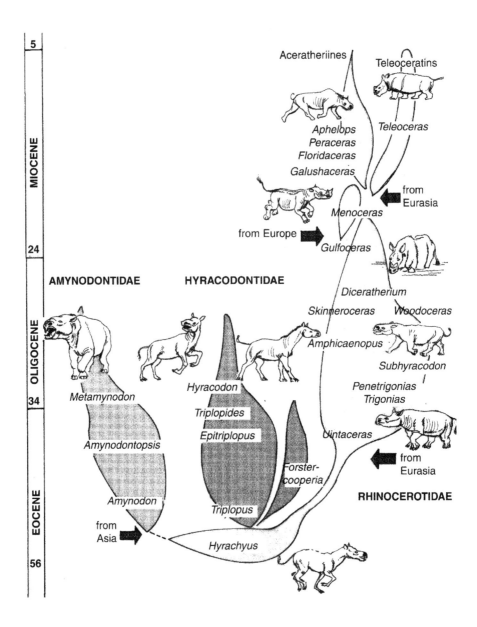

Figure 4.4. Rhino history. (From Prothero, 2005; courtesy C. R. Prothero.)

ear region) and the top of the skull (especially the nasal region) and skeleton (certain features of the limb bones) that help identify a fossil rhino, but the first thing that a good paleontologist notices is the teeth. If they show these characteristic patterns, they belong to a rhino and the paleontologist then needs to ask, "What kind of rhino is it?"

Rhino Radiation

With these basic facts about rhino anatomy, we can trace their history back to the early Eocene, about 50 Ma (Fig. 4.4). The oldest known member of the rhinoceros lineage (the superfamily Rhinocerotoidea) is a creature called *Hyrachyus* (pronounced HY-ra-KY-us), whose name literally means "hyrax pig" in Greek (Fig. 4.5). Joseph Leidy gave it that name in 1871 because its tiny rhino-like teeth resemble those of the living hyrax, or coney, of Africa and Asia, yet its teeth are low crowned with discrete cusps like those of a pig. *Hyrachyus* was about the size of a large dog, yet it had a skull not too different from the earliest horses and earliest tapirs, the other two main groups (along with rhinocerotoids) that make up the odd-toed hoofed mammals, or Perissodactyla. But *Hyrachyus* differs from the earliest horses and tapirs in that it already had the classic rhinocerotoid feature of π-shaped crests on the upper molars and L-shaped crests on the lower molars (Fig. 4.2C). The rest of the skeleton is very lightly built with long legs for running, and three long toes on its front and hind feet (and sometimes a vestigial fourth toe, the fifth metacarpal, on each hand). *Hyrachyus* was very widespread and successful in the late early Eocene, occurring in great numbers in classic North American localities like the Bighorn, Wind River, and Washakie Basins of Wyoming. It even spread to Europe and lived above the Arctic Circle on Ellesmere Island in what is now the Canadian Arctic. This occurred during a period of time when the planet was in a "super-greenhouse" climate so that there was no ice anywhere

Figure 4.5. Hyrachyus, *the most primitive known relative of all the rhinocerotoids. (Restoration by R. Bruce Horsfall; from Scott, 1930.)*

*Figure 4.6. Amynodonts were hippo-like aquatic rhinos. These are the typical amynodonts of the
Oligocene of the Big Badlands,* Metamynodon planifrons. *(Restoration by R. Bruce Horsfall;
from Scott, 1930.)*

(see Chapter 7), and the temperature above the Arctic Circle was warm enough
for alligators and pond turtles and a wide variety of mammals that apparently
traveled across the Greenland-Iceland land bridge between Europe and North
America. *Hyrachyus* was one of the widest roaming of all these mammals.

From such an unremarkable creature as *Hyrachyus*, barely distinguishable from
the earliest contemporary horses and tapirs, rhinos soon diverged into three easily
distinguished families, two of which are now extinct. One family, the amynodonts
(family Amynodontidae), developed into huge hippo-like forms with long flaring
tusks (Fig. 4.6), and some of their kind even developed highly specialized skulls
with retracted nasal openings, suggesting that they had a trunk or proboscis like
a mastodont or tapir. Amynodonts flourished in the middle and late Eocene, but
vanished in North America after the early Oligocene. Their last surviving member, *Cadurcotherium*, lingered on in Asia until 20 Ma and is found in the same
beds in Pakistan (Dera Bugti) that produces indricotheres (see Chapter 3). Their
upper molars (Fig. 4.2A) are very distinctive in that they have a large crest ex-
tending from the outer back part of the tooth (called a metastyle), much larger
than the condition of these crests in *Hyrachyus* or any other family of rhino.
Amynodonts also modify the primitive front teeth of *Hyrachyus* (Fig. 4.7A) and
develop the upper and lower canines into robust tusks (Fig. 4.7E) that typically
flare out of the front and side of the mouth like those of modern hippos. These
features, along with their short limbs, barrel-shaped bodies, and the fact that they
are often found in river channel sandstones, suggest that they may have lived
much like hippos, spending most of their time in the water. However, based on

the low-crowned teeth of amynodonts and the absence of grasses in the Eocene, they must have eaten much softer, leafier vegetation than the modern hippo (*Hippopotamus amphibius*), which is a nocturnal grass eater or grazer.

The second family is the true rhinoceroses (family Rhinocerotidae), which includes the five living species and dozens of species of extinct relatives (Fig. 4.4). They are discussed in much greater length in two of my previous books, so I will not dwell on them here (Prothero and Schoch, 2002; Prothero, 2005). They have many distinctive features, including a last upper molar that loses the metastyle entirely, making its crests form a V pattern rather than the typical rhino π pattern found on the rest of its upper molars. In addition, rhinocerotids develop a tooth combination of a chisel-shaped upper third incisor that occludes against a tusk-like lower incisor (Fig. 4.7F). These lower tusks in rhinos are very important as defensive weapons (much more important than the horns in Indian rhinos, for example) and for sparring with other rhinos. These tusks differ in size and shape

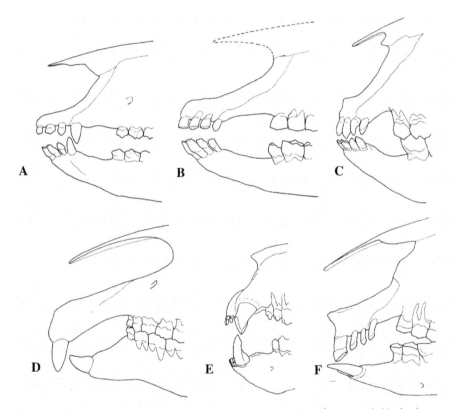

Figure 4.7.The front teeth of rhinocerotoids are diagnostic of family groups. A. Starting with Hyrachyus, *hyracodonts developed spatula-shaped incisors (B,* Triplopus; *C.,* Hyracodon). *D. The front teeth of* Paraceratherium. *E. Amynodonts, on the other hand, developed prominent upper and lower tusks. F. True rhinocerotids, such as* Trigonias, *have a chisel-shaped upper incisor that occludes against a tusk-like lower incisor. The remaining upper incisors are lost in all later rhinos. (From Radinsky, 1966; used with permission.)*

between males and females (males have long curved pointed tusks, while those of females are short and blunt), so the gender of fossil rhino skulls and jaws can be determined.

Hyracodonts and Indricotheres

The third family is the hyracodonts, family Hyracodontidae, the group from which the giant indricotheres arose. From early in their evolution, hyracodonts developed relatively long limbs more specialized for running than any other rhino ever developed, hence their nickname, the "running rhinos." In the middle Eocene, forms like *Triplopus* looked much like *Hyrachyus,* except that they were larger in body size and had characteristic specializations of their teeth. Like true rhinocerotids, hyracodonts have lost the metastyle on their third upper molars (Fig. 4.2C), giving this last tooth the V-shaped crests rather than the π-shaped crests of the rest of their upper molars. Unlike rhinocerotids, however, hyracodonts did not develop the upper chisel-lower tusk combination of third incisors in the front of their mouths. Instead, they had large conical cusps on their third incisors, and their front incisors were shaped like little spatulas (Fig. 4.7C) rather than like pointed cones as in most other mammals. Finally, as discussed below, all hyracodonts had remarkably long feet and toes, making them better adapted for running than their contemporaries.

During the late Eocene and Oligocene, hyracodonts, like the genus *Hyracodon* itself (Fig. 4.8, foreground of Frontispiece), were among the most abundantly

Figure 4.8. The Great Dane-sized running rhinoceros Hyracodon. *(Restoration by R. Bruce Horsfall; from Scott, 1930.)*

preserved mammals in areas like the Big Badlands of South Dakota. Today, we have literally hundreds of good skulls and many good skeletons, so we can talk about their evolution and anatomy in great detail (Prothero, 1996a). *Hyracodon* fossils are easily recognized and very distinctive, with π-shaped upper molars and premolars that are gradually evolving and transforming into the π pattern as the different species evolved (Prothero, 1996a). The teeth are not very complex or high-crowned, however, so *Hyracodon* probably ate soft, leafy vegetation or browsed on scrub; grazing on gritty grasses was not common yet, because grasslands had not yet appeared.

The legs and especially toe bones of *Hyracodon* are even longer proportionally than those of *Triplopus* or *Hyrachyus,* and the larger specimens reached the size of a Great Dane dog. This made it only slightly larger than the common three-toed Badlands horse, *Mesohippus,* and both (along with the humpless camel *Poebrotherium*) were among the fastest runners in the early Oligocene. Its skull, however, was considerably larger than that of contemporary horses, and the neck was longer and more robust, so it did not look like any horse of that time, nor any modern horse, for that matter.

Meanwhile, as *Hyracodon* (a member of the hyracodontine lineage) remained small and became more specialized for running across the plains of North America, a different lineage was evolving at the same time. Starting with the genus *Forstercooperia* from the middle and late Eocene of Eurasia, the descendants of this lineage became larger and larger. (*Forstercooperia* was once reported from North America by Lucas et al., 1981, but those specimens have now been relegated to the true rhinocerotid family by Holbrook and Lucas, 1997.) There is not only *F. jigniensis,* reported from the Eocene of India (Sahni and Khare, 1972), and *F. totadentata, F. grandis*, but also *F. minuta* from the middle Eocene of China and Mongolia (Lucas et al., 1981; Lucas and Sobus, 1989). All of the *Forstercooperia* species are relatively small as indricotheres go, from *F. minuta* (the size of a collie dog) to *F. grandis* (the size of a sheep). They still have relatively primitive skulls without the high degree of nasal retraction (and therefore proboscis) that later indricotheres developed. In addition, they had relatively small front teeth and upper tusks, and a trace of the metastyle on the third upper molars, which is lost completely on later indricotheres (among other features).

By the end of the middle Eocene (end of the Sharamurunian Asian land mammal age), *Forstercooperia* had vanished from Asia. But later forms of indricotheres persisted in Asia, where their evolution really took off. In the late middle Eocene (Sharamurunian), we find specimens of a larger genus, known as *Juxia* (pronounced jü-HSIA). *Juxia borissiaki* and *Juxia sharamurenense* are both recovered from beds in Mongolia and China (Figs. 4.9, 4.10, 4.11). Although cow-sized (about 400–800 kg, according to Qiu and Wang, 2007) and only slightly larger than *Forstercooperia, Juxia* already had evidence of the beginning of retracted nasals for a proboscis (Fig. 4.10), although it did not yet have the down-turned snout seen in later indricotheres. In 2003 B. N. Tiwari found additional

Figure 4.9. The skeleton of the intermediate-sized indricotheres Juxia, *as mounted in the IVPP in Beijing. (Courtesy Deng Tao.)*

Figure 4.10. The skull of Juxia sharamurunense. *(From Qiu and Wang, 2007; used with permission.)*

Figure 4.11. Restoration of Juxia. *(From Qiu and Wang, 2007; used with permission.)*

specimens of *Juxia* from the Liyan Formation of eastern Ladakh in India. Wang (1976) described some Chinese specimens that she referred to her new genus *Imequincisoria*, but Lucas and Sobus (1989) showed that this material is just more fossils of *Juxia*, so that name is not considered valid.

By the late Eocene, the indricothere lineage was getting really large. The upper Eocene Ulan Gochu beds or Urtyn Obo Formation of Mongolia yield fossils of the next member of their lineage, *Urtinotherium incisivum* (Fig. 4.12). First distinguished from specimens originally called *Indricotherium* and given its own generic name by Chow (also spelled as Zhou) and Chiu in 1963, it looks much like the gigantic indricotheres of the Oligocene (*Paraceratherium* and possibly *Indricotherium*), except it was just a bit smaller (Fig. 5.3C). Although it is known mostly from lower jaws and a few isolated upper cheek teeth, it is the oldest indricothere to show the huge conical lower first incisor which points forward, not upward (Fig. 4.13). This forward-pointing, or procumbent, condition of the incisor is a distinctive feature of all later and bigger indricotheres. Unlike most of its descendants, however, *Urtinotherium* still retains the second and third lower incisors, lower canines, and lower first premolars. These teeth were lost in later indricotheres, as they have highly specialized lower jaws with only a single conical procumbent (forward pointing) incisor on each side of the jaw and no other teeth behind the first incisor until the second lower premolar of the cheek tooth row (Fig. 4.13). The long toothless gap between front teeth and cheek teeth is called a diastema, and it is a common feature in many mammals that develop very different specializations between their front incisors and their cheek tooth row.

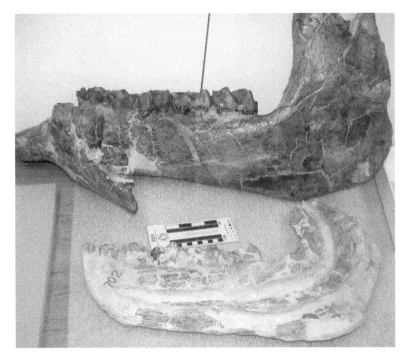

Figure 4.12. Urtinotherium *jaw (foreground) compared to* Paraceratherium *jaw from Mongolia (background). Scale bar in cm (bottom) and inches (top).*

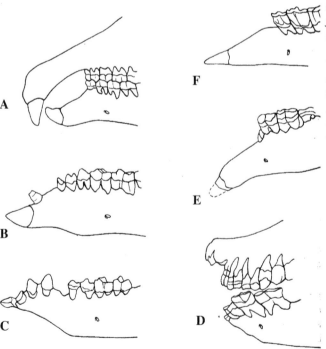

Figure 4.13. Front teeth and snout regions of different indricotheres, as recognized by Qiu and Wang (2007). A. Juxia. *B.* Urtinotherium. *C.* Paraceratherium. *D.* Dzungariotherium. *E.* Aralotherium. *F.* Turpanotherium. *(From Qiu and Wang, 2007; used with permission.)*

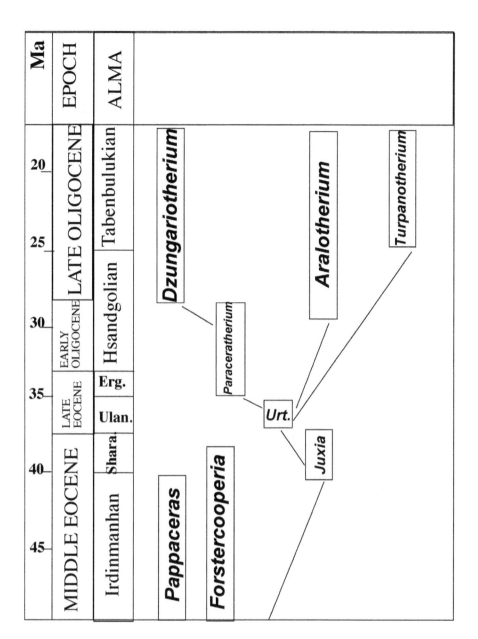

Figure 4.14. Family tree of indricotheres (as interpreted by Qiu and Wang, 2007). Urt. =
Urtinotherium. *ALMA = Asian land mammal ages. Ma = million years ago. Erg. = Ergilian;*
*Ulan. = Ulangochuan. If Lucas and Sobus (1989) are correct, most of the Oligocene taxa (*Aralotherium, Turpanotherium*) are junior synonyms of* Paraceratherium, *and only* Dzungariotherium *might be a distinct taxon. (Modified from Qiu and Wang, 2007.)*

Horses, for example, have a large diastema between their incisors and cheek teeth, and that gap is where a rider puts the bit of the bridle when preparing a horse for riding.

Putting this all together, we can sketch out a family tree of the Asian indricotheres (Fig. 4.14), beginning with the middle Eocene rhinos *Forstercooperia* and *Juxia*, the late Eocene *Urtinotherium,* and the beginning of the giant Oligocene forms.

And finally we come to the most remarkable rhinos of all, the gigantic indricotheres of the Oligocene of Asia. We will look closer at their anatomy in Chapter 6. First, however, we need to lay down the foundation for how we name animals and clear up confusion in the names of these beasts. That is the subject of our next chapter.

5

What's in a Name?

What's in a name? A rose by any other name would smell as sweet.
—William Shakespeare, *Romeo and Juliet*

Systematics and Taxonomy

In previous chapters, we have seen that the names of some these fossil rhinos are confusing, controversial, or unsettled. Before we go further, we need to look at how animals and plants are named and what rules must be followed to see why these disputes arise. The science of classifying is known as *taxonomy* (Greek, "laws of order"); any named grouping of organisms (a species, a genus, etc.) is called a *taxon* (plural, taxa). Deciding how to name a new species and genus may seem to be a highly specialized, legalistic dimension of biology and paleobiology, not nearly as glamorous as ecology or behavior or physiology. But taxonomy is not just naming species, because species and higher taxa reflect evolution. Taxonomists do much more than label dusty jars in a museum. They are interested in comparing different species and deciding how they are related and ultimately in deciphering their evolutionary history. They look at the diversity of organisms in time and space and try to understand the large-scale patterns of nature. They look at the present and past geographic distributions of organisms and try to determine how they got there. In short, they look at the total pattern of natural diversity and try to understand how it came to be. Contrary to stereotypes, they are among the most eclectic of biologists and paleobiologists.

All these various enterprises go beyond conventional taxonomy and are usually given the broader label of *systematics*. Systematics has been defined as "the science of the diversity of organisms" (Mayr, 1969, p. 2) or "the scientific study of the kinds and diversity of organisms and of any and all relationships among them" (Simpson, 1961, p. 7). Its core consists of taxonomy, but it also includes determining evolutionary relationships (*phylogeny*) and determining geographic relationships (*biogeography*). The systematist uses the comparative approach to the diversity of life to understand all patterns and relationships that explain how life came to be the way it is. Put this way, systematics is one of the most exciting and stimulating fields in all of biology and paleobiology.

Taxonomists and systematists may not be as numerous or well funded as molecular biologists or ecologists or physiologists or behaviorists, but their labors are essential. All other disciplines in biology and paleobiology depend upon taxonomists to give their experimental subjects a name and, more importantly, to give them a comparative context. If a physiologist wants to study the organism that is most like humans, it is the taxonomist who points to the chimpanzee, our closest evolutionary relative. If an ecologist wants to understand how a particular symbiotic relationship may have developed, or the ethologist wants to understand a peculiar type of animal behavior, they need to know the evolutionary relationships and phylogenetic history of each organism, and these are the domain of the systematist. Systematics provides the framework of understanding and interconnection upon which all the rest of biology and paleobiology are based. Without it, each organism is a random particle in space, and what we learn about it has no relevance to anything else in the living world.

In our present age, taxonomists have become scarce as grant funding dries up and students go into more glamorous specialties that require big, expensive machines. Yet one of the most important issues on this planet today—biodiversity—is within the domain of systematists. Without someone to describe, name, and count all the species on this planet, how will we know whether we are wiping them out catastrophically, or whether they are holding their own or even flourishing? Without the perspective of past diversity changes on this planet, how can we decide the severity of human-induced mass extinction? Each time someone surveys a patch of rainforest, trying to determine how humans have impacted the life there, his or her first task is taxonomy. Ecologists complain that they cannot find anyone who has the right training to identify and to describe all the new species of insects and birds and plants that are being destroyed even before we get to know them. Without knowing that they are there, how can we decide how important they might be? One of these species might hold the cure to some deadly disease or the solution to the control of a nasty pest, but without systematic and taxonomic research, these species go extinct before we even encounter them.

In the context of paleontology, the situation is analogous. The public may think that collecting big dinosaur specimens in exotic places is exciting, but it is just a tiny part of paleontology. Collecting and preparing fossils is a specialized task, often performed by people with little advanced scientific training. Analyzing and understanding their taxonomy, geography, and phylogenetic relationships is the domain of the systematic paleontologist. Without a properly trained paleontologist to correctly identify, name, and analyze the fossils, they are mute stones. Hours in the laboratory and museum collections spent measuring and describing specimens may not seem as glamorous as visiting exotic places, but they are equally essential. From this naming and description comes the understanding of larger problems in paleobiology, such as: how is all life interrelated? What is the past history of life? How has diversity on this planet changed? Without the foundation of systematics, these questions cannot even be approached.

Rules of the Road

When Linnaeus and other early natural historians developed different schemes of classification, there was no general agreement on how it should be done. Linnaeus' system became so successful that it soon became the standard in most parts of the world, but still there were no official rules, and chaos reigned. If one systematist didn't like a particular name for an organism, he might rename it for no good reason. Another systematist might use a name that had already been used for some other animal. Still another might name the species in his native language or name the species after himself. Some taxa were given more than one name. Systematics became a battleground of natural historians squabbling over proper names, and there were no referees to break up the fights.

To bring order out of this chaos, rules were needed. In 1842 Strickland proposed the first code for zoology. Over the years these codes have evolved, and the first international code of zoological nomenclature was published in 1905. The current International Code of Zoological Nomenclature (ICZN) was last revised in 2000 (available both in printed form and online at http://iczn.org/code).

Taxonomic codes of nomenclature have an important purpose: to enhance stability and improve communication when creating or using taxonomic names and making taxonomic decisions. Systematists around the world are bound to follow these rules if they want their taxonomy to be recognized by other scientists. If they fail to do so, their work may not be published because journal and book editors follow the codes strictly. If systematic descriptions or new species are somehow published but do not follow the rules, they may be corrected by someone else who does follow the rules. At times, it seems that systematics becomes bogged down in legalistic trivia, but the rules are essential if taxonomists want to avoid unnecessary squabbles and wasted or duplicated effort. It is comparable to knowing the rules of the road before you take your driver's test to get your license. The Department of Motor Vehicles doesn't want you behind the wheel on the streets if you don't know the rules that everyone else is following. Similarly, the international community of systematists avoids "collisions" and "mistakes" by following their own internal set of "traffic rules."

Bound in bright green, the most recent edition (2000) of the ICZN runs to 306 pages, covering 90 articles and 86 recommendations, with the first 126 pages in English followed by a separate section in the equivalent French. (Except for the French, most international zoologists use English in international communication and publication.) The arbitrary starting point of the code is 1758, which is when the tenth edition of Linnaeus' *Systema Naturae* was published. Names and taxa proposed before that date are not bound by these rules (but may not be recognized, either). The code is built around several basic principles:

1. *Binomial nomenclature* (Article 5)—These are the basic rules by which genera and species are created, named, and described. Each binomen ("double name" in Latin) must be based on Latin or latinized words from other languages to enhance international communication across language barriers. The latinized binomial is not always based on actual Latin words, but it must still follow the rules of Latin grammar. For example, if the species name is an adjective, it must be in the same gender (masculine, feminine, or neuter) as the genus name that it modifies. (Although few scientists know Latin these days, it is still useful in surprising ways.) The new taxon must be adequately diagnosed, described, illustrated, named, and published in a recognized scientific journal that is widely distributed and available to most systematists. This does not include unpublished dissertations and local newsletters with limited circulation. In addition to a clear definition and description, the author must also indicate the geographic or stratigraphic range of the taxon and list any relevant measurements or statistics. The origin and meaning, or etymology, of the new name, is also usually indicated (although this is not required). You can base names on any word as long as it is properly latinized, except that you cannot name a taxon after yourself. (You can, however, name it after a friend, and have your friend do the same for you with a different species.) Once a name has been used (even if it later proves to be invalid), it can never be used again for another animal.

The criterion of Greek or Latin roots and latinization of names has become more relaxed as fewer and fewer scientists are learning the classical languages. Less than a century ago, a knowledge of Latin and Greek was the standard for all scholars. I feel very fortunate that I took six years of Latin and three years of Greek in high school and college, because this knowledge has given me a great advantage in remembering, spelling, and deciphering taxonomic names. It has also been valuable in helping me to translate century-old paleontology monographs and in enabling me to correctly compose taxonomic names (and to correct the mistakes made by others). Knowledge of Greek and Latin is becoming less important now that much work is being done in China, Japan, Russia, India, Latin America, and other less western European-influenced scientific communities. Consequently, scientists have gotten more and more creative with their names, often erecting names that are silly or hard for others to use. For example, mammalian paleontologist J. Reid Macdonald (1963) gave names based on the Lakota language to a number of specimens recovered from the Lakota Sioux reservation land near the old site of the Wounded Knee Massacre in South Dakota. Most non-Lakotans find them difficult to pronounce or spell. Try wrapping your tongue around *Ekgmowechashala* (iggi-moo-we-CHA-she-la), which means "little cat man" in Lakota. It is a very important specimen of one of the last fossil primates (or possibly a colugo) in North America. In the same paper, Macdonald also named *Kukusepasatanka,* a hippo-like anthracothere, *Sunkahetanka,* a primitive dog, and *Ekgmoiteptecela,* a saber-toothed carnivore. Then there is the transitional fossil between seals and their ancestors known as *Puijila,* which comes from the Inuktitut language of Greenland; you'll need to visit http://nature.ca/puijila/ fb_an_e.cfm to hear the correct pronunciation. In Australia, there are many fossils that have tongue-twisting names with Aboriginal roots, such as *Djalgaringa, Yingabalanaridae, Pilkipildridae, Yalkparidontidaem, Djarthia, Ekaltadeta, Yurlunggur, Namilamadeta, Ngapakaldia,* and *Djaludjiangi yadjana.* Some others include *Culmacanthus* ("culma" is Aboriginal for "spiny fish"), *Barameda* (Aboriginal for "fish trap"), and *Onychodus jandamarrai,* after the Jandamarra Aboriginal freedom fighters. *Barwickia downunda* is named after Australian paleontologist Dick Barwick. *Wakiewakie* is an Australian fossil marsupial, supposedly named from the Australian way of waking up sleepy field crews in the morning.

As Krishtalka (1989) describes it, about a century ago, an entomologist named Kirkaldy got a bit too creative naming different genera of "true bugs," or Hemiptera. He published the names *Peggichisme* (pronounced "peggy-KISS-me") and *Polychisme* for a group of stainer bugs; *Ochisme* and *Dolichisme* for two bedbugs; *Florichisme* for a plant hopper bug; and *Marichisme, Nanichisme,* and *Elachisme* for seed bugs. For leaf hoppers and assassin bugs, Kirkaldy used male names such as *Alchisme, Zanchisme,* and *Isachisme.* In 1912 the Zoological Society of London officially condemned his naming practices, although they could not abolish the names so long as they were valid taxa.

Several websites devoted to weird names (http://www.curioustaxonomy.net/; http://www.neatorama.com/2007/02/19/the-worlds-strangest-dinosaur-names/) list the gamut of odd inspirations, from puns to wordplay to palindromes that read the same way forward and backward. Some of the more clever names include the clams *Abra cadabra* and *Hunkydora,* the beetle *Agra vation,* the snails *Ba humbugi* and *Ittibittium* (related to the larger snail *Bittium*), the flies *Meomyia, Aha ha,* and *Pieza pi,* the wasps *Heerz tooya* and *Verae peculya,* the trilobite *Cindarella,* the Devonian fossil *Gluteus minimus,* the fossil carnivore *Daphoenus* (pronounced Da-FEE-nus) *demilo,* the fossil snake *Montypythonoides,* the extinct lorikeet *Vini vidivici* (which echoes Julius Caesar's famous statement about Gaul: "I came, I saw, I conquered" or "Veni, vidi, vici" in Latin) and the water beetle *Ytu brutus,* and the "Lizard of Aus," the Australian dinosaur *Ozraptor.* After a few too many beers, paleontologist Nicholas Longrich says he named a horned dinosaur *Mojoceratops,* because it had an elaborate heart-shaped frill that might have improved its ability to attract mates. There is a Cretaceous lizard named *Cuttysarkus* (named by Richard Estes because my graduate advisor, Malcolm McKenna, promised him a bottle of his favorite brand of Scotch whisky if Estes found a Cretaceous mammal jaw). Leigh Van Valen named a doglike fossil mammal *Arfia,* and many of his names for archaic hoofed mammals are derived from

The Lord of the Rings and Tolkien mythical figures. The oldest known primate fossil is known as *Purgatorius,* not because the namer had some sort of religious point to make about humans, but because it was found in Purgatory Hill in the Hell Creek beds of Montana (suitably hellish in the summer time with hot temperatures and dangerous slopes). Despite the musty reputation of taxonomists working away in dark museum basements, never let it be said that they have no creativity or sense of humor!

Although taxonomic names sometimes attempt to describe the creature or give some idea of its main features, if the name becomes inappropriate it is still valid so long as no other senior synonyms are known. For example, the earliest known fossil whales were originally mistaken for large marine reptiles and named *Basilosaurus,* or "lizard emperor." Only later did scientists realize the fossils were from primitive whales, which are mammals, not reptiles, but the name is still valid even if it is inappropriate. In the 1920s scientists retrieved material of a bizarre predatory dinosaur from the Cretaceous of Mongolia and named it *Oviraptor* ("egg thief") from its proximity to nests of eggs they thought belonged to the most common dinosaur there, the horned dinosaur *Protoceratops.* But in the 1980s and 1990s, expeditions returned to Mongolia and found fossil skeletons of *Oviraptor* mothers brooding those same eggs, and the bones of unborn *Oviraptor* inside the eggs. The "egg thief" was actually the *parent* of the eggs, not a thief at all—but this slanderous name cannot be changed just because it's now inappropriate.

In addition to names with difficult, odd, or funny pronunciations and meanings, there are also not only names which honor individuals, but also names where people have named a tick or a leech or some other parasite after people they wished to *dishonor.* Even though the ICZN has a clause stating, "No zoologist should propose a name that, to his knowledge, gives offense on any grounds," the rule has been violated many times. Linnaeus himself named a noxious weedy aster *Sigesbeckia* after his rival Johann Sigesbeck, who opposed Linnaeus' sexual classification of plants. A zoologist named a piranha *Rooseveltia natteri* because he hated President Theodore Roosevelt. Three different species of slime mold beetles are named after former President Bush, Vice-President Cheney, and Defense Secretary Donald Rumsfeld. There is a species of louse named after the *Far Side* cartoonist Gary Larson (*Strigiphilus garylarsoni*), although this was intended to honor, not dishonor him (and reportedly Larson loved it). The famous late nineteenth-century paleontologists Edward Drinker Cope and O. C. Marsh insulted each other with naming wars. Marsh named a marine lizard *Mosasaurus copeanus* (emphasis on the last four letters), and Cope named a fossil hoofed mammal *Anisonchus cophater* (emphasis on the last five letters). Cope told his protégé Henry Fairfield Osborn, "Osborn, it's no use looking up the Greek derivation of cophater, . . . for I have named it in honor of the number of Cope-haters who surround me. . . ." A century later in 1978, Leigh Van Valen returned the compliment by naming another primitive hoofed mammal after Cope: *Oxyacodon marshater.* The huge piglike mammal *Dinohyus hollandi* (Fig. 3.6A) was named by paleontologist O. A. Peterson after his museum director W. J. Holland, who put his name as first author on every paper, even if he didn't do the research or write any of it. The name means "Holland's terrible pig." When the specimen was announced by the Pittsburgh newspaper, they ran the front-page headline, "*Dinohyus hollandi,* The World's Biggest Hog!"

2. *The Principle of Priority* (Article 23)—For the sake of stability and simplicity, the first available name proposed (after 1758) for a taxon is the valid name, except under highly unusual circumstances. Problems and conflict usually arise when two different scientists give different names to the same organism because they were unaware of each other's work, or when more than one name is given to the same organism because some scientists name new species based on the most trivial of criteria. Once the valid name is established, all the later names become invalid synonyms, which cannot be used again. The synonyms can be objective (two scientists actually gave different names to the same specimen) or subjective (a later reviser thinks that two species or specimens are the same, and so one is a synonym of the other).

Normally, this synonymy is established early, so when most scientists learn a name, its priority is no longer in question. Occasionally, however, there are problems. If careful library work or web searches show that some obscure scientist gave a different but prior name to a familiar taxon, that long-forgotten name legally has precedence over the much more familiar name. It doesn't matter that this obscure name was poorly described and poorly illustrated in a minor journal that nobody reads. As long as the name does not violate any of the rules, it has priority. As Charles Michener put it, "In other sciences the work of incompetents is merely ignored; in taxonomy, because of priority, it is preserved."

If the overthrow of a well established name causes too much hardship for scientists, there is one final legal recourse: the International Commission of Zoological Nomenclature can suppress the obscure name through use of its plenary powers. To suppress the name, the taxonomist submits a formal application and justification to an international committee of about thirty scientists, who then publish the case, invite commentary, and decide it by majority vote. This procedure has served taxonomists very well. For example, the widely studied protozoan *Tetrahymena* has been mentioned in over fifteen hundred papers published over twenty-seven years using that name. However, there are at least ten technically valid but long-forgotten names that had priority. Because no purpose would be served by resurrecting these obscure names, the Commission voted unanimously to suppress them.

Sometimes the case is not so clear. Take the dinosaur that everyone knows as "*Brontosaurus.*" In 1877, Yale paleontologist O. C. Marsh published two paragraphs without illustrations on a juvenile specimen of a sauropod he called *Apatosaurus ajax.* Two years later, he described another slightly larger, more complete, and more mature specimen from the same beds as *Brontosaurus*. Like most paleontologists of his time, Marsh was a taxonomic "splitter" who created a new taxon on every slightly different fossil he found. By 1903, Elmer Riggs realized they were the same dinosaur, and without fanfare sank the name *Brontosaurus* as a junior synonym of *Apatosaurus*. As far as scientists are concerned, the case is closed—and the name "*Brontosaurus*" cannot be used, except in an informal sense.

However, Marsh's "*Brontosaurus*" was the most complete sauropod specimen then known, and it became a famous museum display. The reconstructions of this mounted skeleton were then copied and were the basis of hundreds of drawings, paintings, book illustrations, and movie monsters—all bearing the scientifically invalid name "*Brontosaurus*." Because children's books and popular movies seldom check the scientific accuracy of their content with scientists, but shamelessly copy older books and movies, the name was perpetuated, even though no paleontologist has taken the name seriously since 1903. In 1989, the U.S. Postal Service made the news when they issued a "*Brontosaurus*" stamp and then received criticism from paleontologists for using an invalid name. Some think that the name *Apatosaurus* should be suppressed, since "*Brontosaurus*" is much more familiar (see Steven Jay Gould's essay, "Bully for *Brontosaurus*.") However, the Commission is unlikely to agree, since the synonymy was established one hundred years ago and professional paleontologists haven't used the invalid name since. It may be obscure to the general public (although more and more children's books and popular books now have it right), but that doesn't matter—it's not obscure as far as scientists are concerned.

Problems with prior synonyms have come up with many of the rhinos we discuss in this book. For example, in 1938 Horace E. Wood II first named a rhino *Cooperia* in honor of Sir Clive Forster Cooper, the first discoverer of indricotheres. Then someone pointed out to him that the name *Cooperia* has already been used by Ransom in 1907 for a roundworm (nematode) and could not be used for another organism. So in 1939, Wood had to publish a short notice replacing the rhino name *Cooperia* with another name, *Forstercooperia,* which has never been previously used for any other animal.

As another example, in 1913 Forster Cooper applied the name *Thaumastotherium* ("amazing beast") to some of the indricothere specimens from Pakistan (Dera Bugti). Shortly after he published this name, his colleague C. W. Andrews pointed out that in 1908, the entomologist Kirkaldy (the same one we mentioned earlier, whose naming practices would be condemned four years later) had used *Thaumastotherium* for a species of true bug, or hemipteran. So a few months later in 1913, Forster Cooper renamed those specimens *Baluchitherium* to replace the name already used for a bug.

In addition, there is often a problem with the fact that many early twentieth-century paleontologists were "splitters," giving a new name to nearly every fossil that was found. Not until the 1940s and 1950s did it become common for paleontologists to think about fossils as a part of a highly variable population and "lump" the different variants together as one species. Forster Cooper was certainly a splitter. The first name he applied to specimens of his Dera Bugti indricotheres was *Paraceratherium,* coined in 1911. The names *Thausmastotherium* and *Baluchitherium* came later in 1913. Modern paleontologists regard this as an example of over splitting, since these fossil bone samples are highly variable yet they were clearly contemporaries. Thus, paleontologists consider *Paraceratherium* to be the only valid taxon from the Dera Bugti samples in Pakistan, while *Baluchitherium* has been considered invalid since Matthew (1931) and Granger and Gregory (1936) pointed out that they came from the same locality and could not be distinguished. Unfortunately, because Osborn and others called their first Mongolian specimens "*Baluchitherium,*" that invalid name has become very popular and is still found in all sorts of popular books and websites, even though no paleontologist has used it for almost eighty years and it has no validity in the scientific community.

Thus, there is a continuing struggle between modern "splitters" (especially some paleontologists in China, who do not always think of paleontological species in a biological way) and "lumpers" (most Western paleontologists, whose species concepts are influenced by biological population concepts accepted throughout Western paleontology since the 1940s and 1950s.) We already saw how Lucas and Sobus (1989) sank Chow's taxon *Imequicisoria* as a junior synonym of *Juxia.* Lucas and Sobus (1989) placed the wide spectrum of names from Russian and Chinese paleontologists into the genus *Paraceratherium* (named by Forster Cooper in 1911), the senior name of all the giant Oligocene indricotheres. These names included not only *Baluchitherium* (Forster Cooper, 1911), but also *Indricotherium* (Borissiak, 1915), *Aralotherium* (Borissiak, 1939), *Pristinotherium* (Birjukov, 1953), *Dzungariotherium* (Chiu, 1973), *Benaratherium* (Gabunia, 1955), and *Caucasotherium* (Vereshchagin, 1960). Lucas and Sobus (1989) argued that all of these huge fossils are variants within a single species, and most Western paleontologists have accepted this judgment. However, in Chinese paleontological literature (e.g., Qiu and Wang, 2007), you still might find them using these names that no Western paleontologist accepts because Chinese concepts of what a species name means in paleontology are very different from that held by the rest of the world's paleontologists.

3. *Principle of Coordination* (Articles 36, 43, and 46)—The names of families, subfamilies, and some higher taxa are based on the name of a genus within that family. It doesn't matter if anyone formally establishes this family name—its proper name is automatically established by the naming of that genus. For example, in the case of the extinct rhinoceros family Hyracodontidae, the genus *Hyracodon* was described in 1856 as the first named of all the genera in the family. Even though no one actually published the name Hyracodontidae in 1856, it is considered established. This is despite the fact that other names for the same fam-

ily of rhinos, such as "Triplopodidae," were published before the name "Hyracodontidae" finally appeared in print.

4. *Homonymy* (Article 52)—Names that have already been used cannot be used again, either in an identical spelling, or even if they are spelled differently but sound the same (a homonym). This helps prevent confusion when the names are spoken rather than written; scientific names are hard enough to keep track of without having two different names that sound the same! Cases of homonymy are rare, but they do crop up now and then. For example, in my own research (Prothero, 1996b), I discovered an example. In sorting out the early evolution of camels, I found that the extinct camel *Protomeryx campester* was named in 1904, and another extinct camel was named *Oxydactylus campestris* in 1909. Later I found that the name "*Protomeryx*" was invalid, and some scientists thought that the species *"P." campester* should be included in *Oxydactylus*. But that creates a problem: the genus *Oxydactylus* already had a species *campestris* (both species names would be spelled the same to agree with the gender of the genus). If this homonymy had actually occurred, a different name would be required for one of these camels. Fortunately, a determination of priority by the principle of first reviser showed that "*campester*" was not valid. The correct trivial name for the specimens originally called *"P." campester* was *cedrensis*—so the dilemma was avoided (Prothero, 1996b).

5. *Type Specimens* (Article 61)—When a taxonomist names a species, he or she must also designate one specimen as a standard of reference to represent his or her concept of that species. That specimen is known as the type specimen, or holotype. Typically, it is the best available specimen of the species, and in the original publication it must be clearly illustrated and indicated by museum catalog number. Normally, type specimens must be deposited in a major museum or other reference collection with scientific access so other scientists can examine them. Museums put a special label on their type specimens and often store them in a special place.

Although this idea recalls the typological concept of species prevalent before Darwin, it is really a practical matter. Telling similar species apart is hard enough without having vague definitions. If the original author picks a type specimen, it is possible for later scientists to determine his or her concept of that particular species, even if the original descriptions or diagnoses or illustrations are inadequate. In addition to the type specimen, most taxonomists indicate a number of additional specimens that are considered referable to that species; these are known as the referred material, or hypodigm. This gives later scientists a chance to see the original author's concept of the range of variation within the species. Instead of a single type specimen, some scientists prefer to name several syntypes, including the holotype and several paratypes, to give an idea of this spectrum of variation.

Sometimes, problems emerge with the original type specimen. If the original author did not designate which specimen among the syntypes is the holotype, then a later scientist can do so. This specimen is a later designated type, or lectotype. If the original type specimen is lost, a later reviser is obligated to name a new type specimen, or neotype. If a taxonomist decides that the type specimen is inadequate to tell if the species is really distinct, he or she may sink the species name as a *nomen dubium* ("doubtful name"). If a taxonomist describes a species without following the rules properly (usually because he or she left out a diagnosis, a description, a type specimen, or an illustration), the invalid name is dropped as a *nomen nudum* ("naked name").

By now, most people's heads are reeling with all this jargon and legalese. After all, most people read about biology or paleontology because they like organisms or fossils, not law books. There are systematists who act like lawyers, spending all their time ferreting out obscure names and correcting other people's mistakes in print. However, most systematists regard the Code as a necessary skill to be mastered, just as most drivers learn the rules of the road before they drive. Although relatively few biologists will ever describe a new species or need to know the Code well, most paleontologists find it necessary to do at least some of their own systematics, and many describe more than one new species during their careers.

How Many Different Kinds of Giants Were There?

These previous confusions about rhino names like *Cooperia* or *Thaumastotherium* were easily resolved because these names were clearly junior synonyms (*objective synonyms*) of names given to other creatures, such as nematodes or bugs, because of rules of priority. The synonymy of *Baluchitherium* (Forster Cooper, 1911), *Aralotherium* (Borissiak, 1939), *Dzungariotherium* (Chiu, 1973), *Pristinotherium* (Birkjukov, 1953), *Benaratherium* (Gabunia, 1955), and *Caucasotherium* (Vereshchagin, 1960), by Lucas and Sobus (1989) requires a subjective judgment that these genera are all the same thing. Their argument has been accepted by most Western paleontologists, although not by all Chinese paleontologists, such as Qiu and Wang (2007). The McKenna and Bell (1997) classification took a middle position on this issue, regarding some of these genera as valid but others as synonyms of *Paraceratherium,* without justifying any of their decisions.

So are *all* the Oligocene indricotheres a single variable genus? If that is so, then *Paraceratherium* (Forster Cooper, 1911) from Dera Bugti is the senior name for all of these creatures, and the only valid name. But some paleontologists argue that *Indricotherium,* which was described by Russian paleontologist A. A. Borissiak in 1915 from the Turgai locality north of the Aral Sea region of Kazakhstan, is a different creature than the Pakistani *Paraceratherium* or the Mongolian specimens and deserves to be recognized as a second valid genus.

Qiu and Wang (2007) also argued for the distinctiveness of *Dzungariotherium, Aralotherium,* and their new genus *Turpanotherium.* They gave a number of anatomical characters that supposedly distinguish these genera. As we shall see below, however, the crucial factors are those of size and the snout and front teeth. Most of the other "diagnostic features" listed by Qiu and Wang in the skull region seem to be highly variable and subject to distortion and other post-mortem deformation and breakage on the bones. About the only valid reasons for distinguishing them would be if their sizes are distinctly different from typical *Paraceratherium* or if they have unique combinations of teeth or snout features. Based on the diagram of these snout features shown in Fig. 4.13, only *Dzungariotherium* has a unique combination of front teeth. On the other hand, *Aralotherium* and *Turpanotherium* have the typical upper and lower jaw condition of *Urtinotherium* and *Paraceratherium*—large conical lower first incisors that are procumbent (to various degrees, possibly influenced by post-mortem distortion). We will consider the issues of size below.

Lucas and Sobus (1989) argued that all these animals belong in the same highly variable genus (*Paraceratherium*) based on several criteria. First, they noted that in their description of the Mongolian material called "*Baluchitherium*," Granger and Gregory (1936, pp. 54–62) originally made an argument that they were all the same:

> As to generic relationships, our material indicates that both *Baluchitherium* and Indricotherium are close to or even synonymous with *Paraceratherium*, although possibly representing a different species. In the first place, the type upper molars of *Paraceratherium bugtiense* Pilgrim exhibit no conspicuous differences from those of *Baluchitherium* osborni; secondly, the lower jaw referred to *Paraceratherium bugtiense* by Forster Cooper seems to be indistinguishable in generic characters from one of our jaws (AMNH No. 26166) that is associated with our humerus, radius, ulna, and metacarpal III of the general size and characters of Borissiak's *Indricotherium*; thirdly, the cast of the skull refereed by Forster Cooper to skulls of *Paraceratherium* reveals essential similarities at all point to our large skulls of *Baluchitherium* grangeri Osborn. . . . Fourthly, the peculiar lower front teeth of *Paraceratherium* are matched precisely in *Baluchitherium osborni* and in Borissiak's *Indricotherium*. Fifthly, we have numerous fully adult limb bones, astragali and metapodials that collectively comprise a closely graded series . . . from the small *Paraceratherium* through *Baluchitherium* osborni to *B. grangeri* and finally to a super-*Indricotherium*. On the other hand, Borissiak has pointed out that in Forster Cooper's *Paraceratherium* the protoloph of the fourth upper molar is higher than in *Indricotherium*, the whole crown is slightly more hypsodont and the cingulum better developed; also the incipient "crochets" of the upper

molars are a little more pronounced; assuredly, however, the evidence assembled . . . is not favorable to the idea that *Paraceratherium*, *Baluchitherium*, and *Indricotherium* are distinct genera, although there are minor and perhaps specific differences in the second upper premolars.

This statement lays out the case for synonymy pretty clearly, and it is remarkable that Granger and Gregory (1936) did not follow their own arguments to their logical conclusion and at least sink *Baluchitherium* as a junior synonym of *Paraceratherium*. From what we know of the politics at the American Museum in the 1930s, they probably didn't do so because their boss, Henry Fairfield Osborn (a notorious hyper-splitter), would probably not have been happy to see a genus he'd discussed frequently and heavily publicized disappear into synonymy (and with it the species named after him, *Baluchitherium osborni*).

Lucas and Sobus (1989, p. 371) point out that most of the dental differences mentioned by many authors, especially those in the upper premolar crests, no longer have any meaning to distinguish and diagnose genera, since they are highly variable within a single population (Gregory and Cook, 1928; Prothero, 1996a, 2005). At best, they might be used to tell the different geographic species apart. Lucas and Sobus (1989) point out that the argument boils down to the shapes of two different types of skulls (see Fig. 5.1):

1. Skulls termed *Paraceratherium* and *Dzungariotherium* have relatively slender maxillaries-premaxillaries; nearly horizontal zygomata; shallow skull roofs above the orbits; relatively thin and posteriorly placed mastoid-paroccipital processes (so that the external auditory meati are relatively wide); less posteriorly extended lambdoid crests and horizontally oriented occipital condyle.

2. The single skull assigned to *Indricotherium* has robust maxillaries-premaxillaries; upturned zygomata; domed frontals above the orbits, thick mastoid-paroccipital processes (so that the external auditory meati are antero-posteriorly constricted); a posteriorly extended lambdoid crest and vertically oriented occipital condyles.

Thus, Lucas and Sobus (1989) follow Granger and Gregory (1936) in arguing that *Paraceratherium*, *Indricotherium*, and the rest all belong in the same genus, with the large domed-skull forms that have been called *Indricotherium* possibly representing males (the single American Museum skull from Mongolia is the only example, see Fig. 6.1), and the flatter-skulled *Paraceratherium* (or *Baluchitherium*, *Aralotherium*), exemplified by most of the skulls recovered from Dera Bugti, representing females (Fig. 5.1). To bolster their argument, Lucas and Sobus (1989, Fig. 19.2) plotted dimensions of the teeth of these rhinos, and the *Indricotherium* and *Paraceratherium* and other large genera of indricotheres cannot be separated based on their plots. Indeed, if you place tooth rows of Dera Bugti *Paraceratherium* side-by-side with Mongolian specimens (once called "*Indri-*

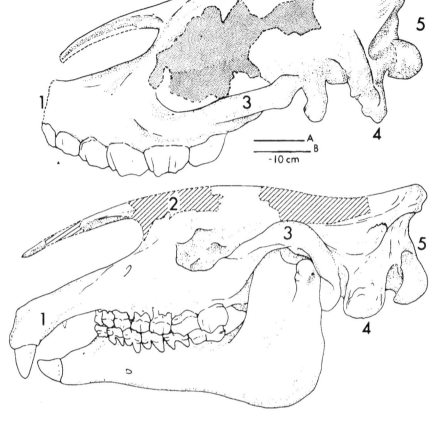

Figure 5.1. Contrast between skulls labeled "Indricotherium" *(bottom) and* Paraceratherium *(top). Features used to distinguish them: 1: robust premaxillaries; 2: domed frontals above orbits; 3: upturned zygomatic arches; 4: massive mastoid-paroccipital processes; 5: orientation of occipital condyles and lambdoid crest. (From Lucas and Sobus, 1989; courtesy S. Lucas.)*

cotherium" or "*Baluchitherium*"), there are no obvious differences, and they are extremely similar in size and shape (Fig. 5.2).

   To evaluate this argument, I remeasured the available specimens in the AMNH and retrieved the original tooth size data from the literature and replotted most of the same teeth originally plotted by Lucas and Sobus (1989), with additional new data from the Qiu and Wang (2007) monograph. These data are shown in Figure 5.3. As can be seen in these plots, the small genera (*Forstercooperia* and *Juxia*) form a distinct small size cluster and are overlapping in tooth dimensions, and *Urtinotherium* (Fig. 5.3C) is just slightly smaller than the giant indricotheres. More importantly, the Mongolian and Chinese and Kazakh specimens called "*Indricotherium*," "*Aralotherium*," "*Turpanotherium*," and other large genera synonymized by Lucas and Sobus (1989) form a single discrete size cluster with no outliers or sub-clusters that would justify these additional genera. In addition, the specimens of *Paraceratherium* from Dera Bugti plot entirely within the range of

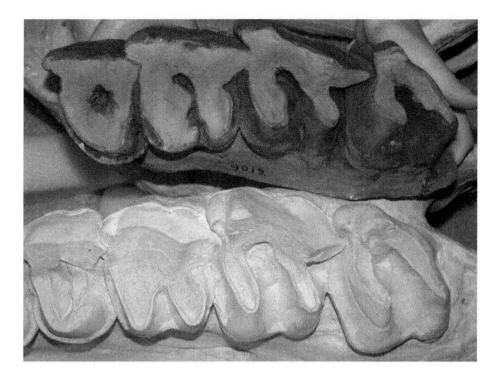

Figure 5.2. Comparison of Dera Bugti Paraceratherium *left upper fourth premolar to third molars (top) with Mongolian* "Indricotherium" *upper teeth, showing that their size and shape are virtually identical and cannot be distinguished visually or by size. Hand in the upper right corner of the photo shows the immense size of these teeth. (Photo by the author.)*

these Chinese-Mongolian-Kazakh specimens, showing that they are indistinguishable in tooth size.

Looking closer at the lower first molar dimensions (Fig. 5.3C), it is clear that the Mongolian and Dera Bugti *Paraceratherium* and *Aralotherium* form a single cluster with complete overlap. *Dzungariotherium*, with its distinctive front teeth (Fig. 4.13) tends to be on the large end of this cluster, so it is potentially distinguishable on these features. Only Qiu and Wang's (2007) taxon *Turpanotherium* is distinct from this cluster, but seems to group with *Urtinotherium*. Thus, I will regard *Turpanotherium* as a likely junior synonym of *Urtinotherium* until further study is conducted.

Yet another argument can be made about this size data. Taking the entire large indricothere data set and analyzing it statistically, we can calculate what is known as the "coefficient of variation" (CV) of each tooth dimension by dividing the standard deviation of the measurement by its mean, then multiplying by one hundred (Simpson et al., 1960; Lande, 1977; Sokal and Rohlf, 1994; Polly, 1998; Plavcan and Cope, 2001; Dayan et al., 2002; Meiri et al., 2005). Fifty years of zoological research indicates that for most mammal populations, CVs of most parts of the anatomy will be less than ten except for features that are extremely sexually dimorphic (Kurtén, 1953; Simpson et al., 1960; Yablokov, 1974).

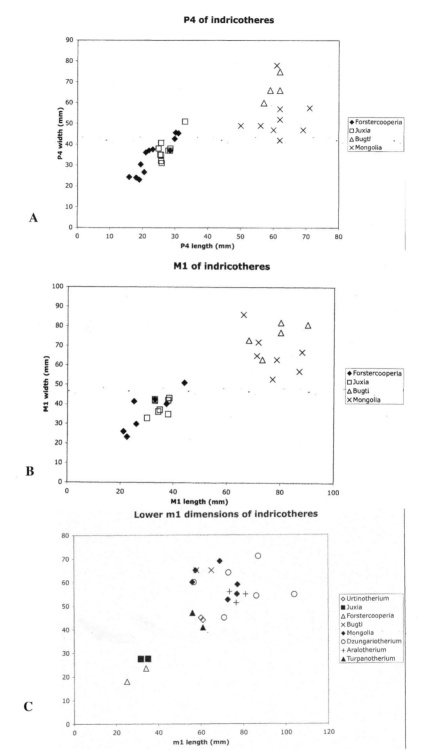

Figure 5.3. Plots of tooth dimensions of indricotheres. A. Last upper premolar (P4). B. First upper molar (M1). C. First lower molar (m1).

Table 5.1. Statistics of Indricothere Teeth

	P4L	P4W	M1L	M1W	m1L	m1W
Mean	68.7	58.5	77.5	70.5	63.8	66.6
Standard Deviation	5.5	11.3	7.6	10.1	6.7	4.5
Coefficient of Variation	8.9	19.3	9.7	14.3	10.5	6.8

In the case of the large indricotheres, the CVs (Table 5.1) are medium to large (in the range of six to nine), but only one dimension gives a CV larger than ten. Moreover, there are no obvious subclusters of specimens (Fig. 5.3) among the data from Dera Bugti, Mongolia, China, and other Asian localities. Thus, it is impossible to make a statistical argument that these allegedly different genera are truly different in size, since their variability can be better explained as coming from a single variable population.

Lucas and Sobus (1989, p. 372) further support the single-genus argument by pointing out that there is a small population sample of skulls from the Turpan Basin, Xinjiang, China, originally described by Xu and Wang (1978). Xu and Wang (1978) had placed these specimens into two species, *Paraceratherium lipidus* and *Dzungariotherium turfanensis,* since they represent the two different skull shapes already mentioned. Yet the fact they come from a single locality suggests they are all one population and should be in the same highly variable species, just like the co-occurrence of the *Paraceratherium* and *"Baluchitherium"* specimens in Dera Bugti suggests that these two genera are synonyms.

Since these arguments were first published in 1989, most paleontologists seemed to have concurred, and no one has published a detailed counter-argument against Lucas and Sobus (1989). Even though some Chinese authors (e.g., Qiu and Wang, 2007) recognize multiple generic names and many different species, this practice is not followed by most Western scientists. Fortelius and Kappelman (1993) used the name *"Indricotherium"* and recognized that *"Baluchitherium"* was invalid, but did not even cite Lucas and Sobus (1989) nor did they mention any of the evidence that *Paraceratherium* was the senior synonym of all the indricotheres. McKenna and Bell (1997, p. 478) listed both *Paraceratherium* and *Indricotherium* as valid genera (along with a number of Chinese and Russian taxa no longer considered distinct) but made no effort to address the arguments of Lucas and Sobus (1989) or present an argument for the distinctiveness of all these genera.

So where do we stand? Certainly, it is revealing that those paleontologists most actively involved in current indricothere discovery and research (P.-O. Antoine, J.-L. Welcomme, G. Métais, and others in the French paleontological community) accept the argument that these fossils are all *Paraceratherium,* even as they discover new specimens in Turkey, Pakistan, and elsewhere. The evidence from the size of the teeth (Figs. 5.2, 5.3) seems compelling that these creatures could not be distinguished by size. As Lucas and Sobus (1989) and our data above showed, the Chinese population sample is also strong evidence that the many different "genera" are the same.

There is another argument that has yet to be considered. The home range, or territory, of a mammal is directly related to its body size, so each species of indricothere must had a home range that was larger than that of any living mammal. Based on modern examples, large-bodied land mammals require large areas to find enough resources to survive. Lowland gorillas have home ranges of roughly 100 square km. Giraffes require individual ranges of about 280 square km. As du Toit (1990) showed, for large-bodied African mammals, the home range (A_{hr}) scales by body mass (M) in the following formula: $A_{hr} = 0.024\ M^{1.38}$. Thus, indricotheres would have had home ranges of at least 1000 square km and maybe much more if their desert scrub habitats (see Chapter 7) had only limited trees and other resources. There would not have been enough room in Asia in the Oligocene to support more than a few populations of them, let alone many species and genera.

In addition, if these large Oligocene indricothere specimens are all approximately the same size, it is difficult to imagine that such huge creatures capable of roaming such large distances across Asia (i.e., large home range) belonged to several different genera. In modern ecology, when animals of similar size and ecology come into contact, they tend not to live in the same area but subdivide the region or its resources. This is called competitive exclusion. Among rhinos, the African black and white rhinos overlap in the East and South African savannah, but only because the black rhino is a leaf-eating browser and the white rhino is a grass-eating grazer, so they subdivide the niche by different dietary needs. This pattern of a grazing and browsing rhino living together in pairs throughout the Miocene of North America, Africa, and Eurasia was very typical of the group (Prothero et al., 1989; Prothero, 2005), but there is always a clear difference between the presumed browser and grazer in teeth and anatomy. Oligocene indricotheres, on the other hand, had nearly identical body size, teeth, and snouts, so there is no basis for thinking they had different dietary preferences or any other obvious ecological niche differences.

This idea certainly would apply to gigantic indricotheres that apparently roamed freely from Turkey to Mongolia and China. As we have already seen, there is no real geographic difference or separation in size or anatomy of the specimens from Dera Bugti, Kazakhstan, China, or Mongolia, so it seems very unlikely that there were multiple huge genera of indricotheres with large overlapping ranges living side-by-side in the same localities through most of the Oligocene.

As a counter-argument, several reviewers pointed out examples of huge land vertebrates living side-by-side at the same time. In the Miocene of Europe, there were several large genera of proboscideans (including deinotheres almost as big as indricotheres), and these are dramatically different in their anatomy and belong to different genera (Calandra et al., 2008). However, these animals were all specialized in very different ways for different diets, so (like the rhino example above), it is possible to imagine multiple large proboscidean species living contemporaneously in the rich forest/parkland habitat of the Miocene of Europe. It is much harder to imagine indricotheres (which are nearly all identical in size and in most of their anatomy) subdividing the much more limited resource base of the desert scrublands of central Asia in the Oligocene the same way.

Others have pointed out that during the Late Jurassic there were multiple genera of huge sauropod dinosaurs living in the same region, as preserved in units such as the Morrison Formation (Farlow et al., 2010; Sander et al., 2011). Here, the comparison is more problematic, since large sauropods are unlike any large land mammal that has ever lived (in size as well as many anatomical as well as physiological features), and so it is hard to analogize the Late Jurassic forests of the Morrison with the Oligocene scrublands and deserts of central Asia. How so many large herbivores lived together in the Late Jurassic is a problem of understanding a very different type of ecology that may not have any modern counterpart (Farlow et al., 2010).

Thus, I can find no strong argument for recognizing any large indricothere genus except *Paraceratherium* (and possibly a second genus and species, *Dzungaritherium orgosensis*) and follow Lucas and Sobus' synonymy of nearly all of the genera of the large middle and late Oligocene indricotheres. Assuming that these creatures are all *Paraceratherium*, then Lucas and Sobus (1989) recognized and distinguished the following valid species:

1. *Paraceratherium bugtiense* (Pilgrim, 1908) from the upper Oligocene beds of Dera Bugti in Pakistan is the type species of *Paraceratherium*. *Baluchitherium osborni* (Forster Cooper, 1913a) becomes a junior synonym. There is additional newly discovered material of *P. bugtiense* recently described by a French-Pakistani team in Dera Bugti (Antoine et al., 2004; Métais et al., 2009). *Paraceratherium zhajremensis* (Osborn, 1923) consists of specimens from the middle and late Oligocene of India, and it almost certainly a junior synonym of *P. bugtiense* (P.-O. Antoine, pers. comm., 2011).

2. *Paraceratherium transouralicum* (Pavlova, 1922) becomes the senior synonym of the material once referred to *Indricotherium transouralicum* (the domed-skull form, based on a possible male skull from Mongolia). Most reconstructions are based on this species, because it is known from nearly complete skeletal material from the middle and late Oligocene of Kazakhstan, Mongolia, and Nei Monggol in northern China. According to Lucas and Sobus (1989), the following species are synonyms: *Baluchitherium grangeri* (Osborn, 1923), *Indricotherium asiaticum* (Borissiak, 1923), *Indricotherium minus* (Borissiak, 1923).

3. *Paraceratherium* (or possibly *Dzungariotherium*) *orgosensis* (Chiu in 1973) is the largest species of indricothere known, although it is known primarily from the teeth, some of which are at least 25 percent bigger than *P. transouralicum*.

The specimens that show the anterior dentition (Fig. 4.13D) might suggest that it could be generically distinct from *Paraceratherium,* but this hypothesis needs to be evaluated. This species comes from the middle and late Oligocene beds of Xinjiang, northwest China. There are three synonyms: *Dzungariotherium orgosensis* (Chiu, 1973) and (each of the following named after a separate skull) *Dzungariotherium turfanensis* (Xu and Wang, 1978) and *Paraceratherium lipidus* (Xu and Wang, 1978). Although there is variation in skull proportions, perhaps due to sexual dimorphism, all of these "species" are known from a restricted geographical range. They also have distinctive crochets on the first and second upper molars. Arguing against the distinctiveness of this "species" is the fact that skulls referred to *P. lipidus* and *D. turfanensis* are found in the same Chinese quarry (Xu and Wang, 1978), and thus are likely from the same population, making *Dzungariotherium* a junior synonym of *Paraceratherium.*

4. *Paraceratherium prohorovi* (Borissiak, 1939) is based on fossils from the late Oligocene of eastern Kazakhstan. This species consists of very incomplete material, so it is not really resolved whether it can be distinguished from the first three, more complete and easily distinguished species.

Therefore, in this book, we will use the name *Paraceratherium,* keeping in mind that there is some disagreement as to whether the name *Indricotherium* or *Dzungariotherium* can be used for the material from the Turgai beds and Mongolia.

6

Building a Giant

Dry Bones

Now that we have thoroughly explored the geology of the regions that produce indricothere fossils and their evolutionary roots within the rhinocerotoids, let us look more closely at the monsters that still hold the record for the largest land mammal that ever lived.

The most impressive part of the animal is the business end: its immense skull (Figs. 5.1B, 6.1). The nearly complete presumed male skull from Mongolia (American Museum of Natural History, or AMNH 18650) is a truly awesome sight all by itself. It measures about 1.3 meters (52 inches, or almost 5 feet) long, according to Granger and Gregory (1936), 33 cm tall at the back of the skull, 33 cm wide at the base of the back of the skull, and is about 61 cm (about 2 feet) wide across the skull at the zygomatic arches. Another partial skull from Mongolia, AMNH 26165, is even larger than this specimen, measuring about 35 cm wide at the base of the back of the skull and 35.5 cm tall at the back of the skull. An even larger partial skull from Mongolia, AMNH 26167, measures 36.5 cm at the back of the skull base and 38 cm tall at the back of the skull. The presumed female skulls from Dera Bugti (Fig. 5.1A) are almost as large in most of these dimensions (according to measurements in Forster Cooper, 1923a, 1923b), so far as can be determined from their less complete preservation and subsequent deformation and distortion.

The next thing that you notice about the skull (Fig. 6.1), besides its immense size, is the large pair of conical upper tusks that point downward and the pair of short conical lower tusks that point forward. As mentioned in Chapter 4 (Figs. 4.7, 4.13), these teeth are unique among all known rhinos and only occur in *Urtinotherium* and *Paraceratherium* (with all its junior synonyms). These tusks occur at the end of an elongated snout, with a large gap (diastema) between them and the cheek tooth row. Unlike most other groups of primitive rhinos, indricotheres have lost all their remaining front incisors as well as the canines that would normally erupt right behind the incisor tusks (Figs. 4.7, 4.13). Osborn (1923a, p. 6) argued that these tusks were mainly for defensive purposes, but

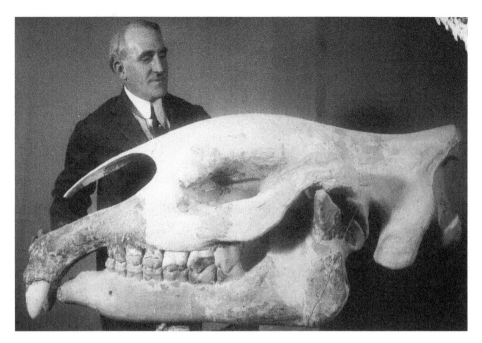

Figure 6.1. Skull usually called "Indricotherium" *(originally called* "Baluchitherium grangeri," *but now* Paraceratherium transouralicum*) from the Hsanda Gol Formation, Mongolia (AMNH 18650), showing its immense size compared to preparator Otto Falkenbach, who reconstructed it from many fragmentary parts. (Image number 310387; courtesy American Museum of Natural History Library.)*

Granger and Gregory (1936, p. 2) suggested that "their primary function was to assist in the sudden jerking loose of shrubs by downward movements of the head and neck, since they are well placed to act thus as picks and levers, while the skull is braced to resist such stresses through its strong rostrum, down-curved zygomata and greatly emphasized basi-occipital eminence."

The next most striking feature is the top of the skull, with the long, smooth domed forehead and no trace in any specimen of a roughened area that would serve as the attachment point for a horn. The bones above the nasal region are long and delicate, and the opening for the nasal incision goes far back into the skull. This enlarged nasal incision is usually a mark of some kind of trunk or proboscis in living mammals, such as tapirs and elephants. Most reconstructions of indricotheres ignore this feature, but they should show a very long prehensile lip, as occurs in living black and Indian rhinos with a shallower nasal incision, and possibly even a short proboscis as well. Such a proboscis is extremely useful to leaf-eaters (browsers) like tapirs, which use them to wrap around branches as they strip off the leaves with their front teeth. Figure 6.2 shows the head muscles and snout-lips reconstruction of *Juxia* by Qiu and Wang (2007). This creature was much smaller with a much smaller nasal incision, so it has a normal rhino upper lip. A large indricothere would have had an even longer, more proboscis-like snout.

Putshkov and Kulczicki (1995) and Gromova (1959) speculated that indricothere tusks were primarily used to break twigs and strip bark, as well as to bend higher branches. They also suggest that early Oligocene specimens (which they call *Indricotherium*) had larger upper incisors, but that later specimens (which they call *Paraceratherium*) had no upper incisors and larger lower incisor tusks, which might have suggested a less leaf-eating and more bark-eating type of diet. However, if Lucas and Sobus (1989) are right, these animals are all in the same genus, and the alleged difference in tusks may be due to Dera Bugti *Paraceratherium* specimens being largely female skulls, while the sole "*Indricotherium*" skull (Fig. 6.1) from Mongolia may come from a male individual. In addition, the new geologic dating of the Dera Bugti beds (Chapter 4) shows that *Paraceratherium* and "*Indricotherium*" are not earlier or later than one another, but contemporaries through most of the Oligocene.

Granger and Gregory (1936, p. 3) noted how low and narrow the back of the skull was, lacking the huge crests across the top (lambdoid crests) and along the midline (sagittal crests) seen in other large animals like brontotheres and elephants. However, both of those mammals had large horns or tusks on the front of the skull, so they would have needed much stronger muscles to support the skull when using their tusks or horns to push or to fight.

At the top of the back of the skull is a deep pit for the attachment of the nuchal ligament, the thick strong ligament that attaches to the neck vertebrae and holds the skull up automatically against the pull of gravity. To make the head bend down, the neck muscles must pull against this "rubber band" ligament that works to hold the skull up without extra muscle action. When land vertebrates die, this ligament tends to contract and pull the neck backwards in a curve, a characteristic pose for many dead animals and fossil skeletons as well.

Even though the back of the skull is low and narrow, Granger and Gregory

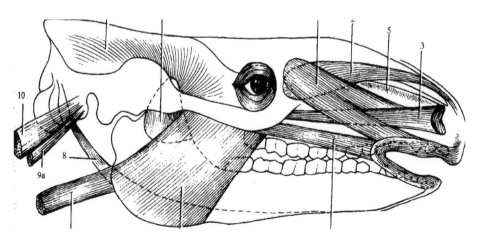

Figure 6.2. Reconstruction of the muscles of the head of Juxia. *(After Qiu and Wang, 2007; used with permission.)*

(1936, p. 3) pointed out that the occipital condyles (the ball joints that surround the spinal cord and connect the skull to the first neck vertebra) are very wide, the paroccipital and mastoid processes are large and robust, and the channels for muscles along the base of the neck vertebrae are also very wide. They conclude that these animals had huge, strong neck muscles that connected all the way down to the front limbs, which would have been more than sufficient to support not only the huge head, but would also allow it to make strong downward sweeps of its head as it stripped vegetation from branches.

One more thing that is not often mentioned: the size and shape of the ears. As we will see below, indricotheres were larger in body mass than any living elephant and almost certainly had problems regulating their body heat at such large size. Elephants must do all they can to increase the surface area of their bodies to release as much excess heat as possible, which is why they have huge fan-like ears full of blood vessels that are essentially giant radiators. Given the huge size of indricotheres, it seems likely that they too should have had elephant-like ears, or at least very large ears of some shape, much larger than they are usually drawn. Most reconstructions try to draw indricotheres as scaled-up modern rhinos with relatively small ears, but we must look to elephants for a different model. Although the soft tissue of the external ear (the "pinna") almost never fossilizes (it is only known in mummified Ice Age mammoths, bison and woolly rhinos), the robust bones around the ear opening (paroccipital processes and mastoid processes) are much like those in some elephants and mastodonts, suggesting that indricotheres should be drawn with much larger ears (as in our cover reconstruction by Carl Buell).

The cheek tooth pattern has already been discussed in Chapter 4 (Fig. 4.2). The upper molars all have the characteristic pi (π) pattern, except for the V-shaped pattern of the third upper molar and the reduced metastyle, a hallmark of hyracodonts and rhinocerotids. The premolars are only partially molarized to form the pattern, but as Gregory and Cook (1928) and Prothero (1996a, 2005) showed, this pattern of molarization of premolars varies within population samples of hyracodonts and rhinocerotids and cannot be used for distinguishing species. These subtle differences in premolar crests are the rationalization behind many of the invalid genera and species of indricotheres erected by Russian and Chinese paleontologists, but as Lucas and Sobus (1989) point out, such features have no systematic value and can no longer be used to diagnose taxa of rhinos. The lower cheek teeth (Fig. 4.3) all have the typical rhino L-shaped pattern, and are highly stereotyped and non-diagnostic, as they are in almost all rhinocerotoids.

In addition to the details of the crown pattern, the teeth are impressive in their sheer size. A single upper molar (Fig. 5.2) is as big as a human fist and larger than almost any other mammal tooth you'll ever see, except for the huge molars of mastodonts, mammoths, and elephants. Yet Granger and Gregory (1936, p. 2–3) note that the size of the molars is relatively small compared to the huge skull. There is also one other important aspect of the teeth: they are relatively simple

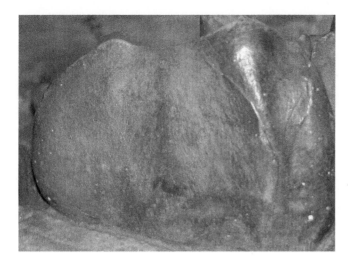

Figure 6.3. Photograph of the mesowear profile of an upper molar of Mongolian Paraceratherium *(looking at the tooth from the external side). The broadly curved convex profile of the top of the molar crest is characteristic of leaf-eating browsers with a small amount of gritty shrubs in their diet. (Photo by the author.)*

and low-crowned, not the high-crowned teeth one finds on later rhinos that have adopted diets of gritty grasses (grazing) that wears down teeth rapidly. Such low-crowned teeth would be indicative of a diet of relatively soft leaves and shrubs (browsing), like most primitive rhinocerotoids.

In recent years, paleontologists and geochemists have developed new techniques for determining the diets of extinct animals, from looking at the sharpness of the tooth crests and vertical relief on the tooth surface (*mesowear*) (Fortelius and Solounias, 2000) to using a microscope to analyze the scratches and pits on the surface of the tooth enamel (*microwear*) (Solounias and Semprebon, 2002). Dr. Matthew Mihlbachler and I both looked at the specimens in the American Museum, and it is clear that indricotheres have mesowear consistent with a leafy, non-abrasive diet, just as predicted (Fig. 6.3). So far as I know, no one has performed an analysis of microwear, but it should show a pattern with few pits or scratches if the indricothere diet was mostly soft leaves (unless their habitat was full of grit and dust, in which case they might have lots of scratches).

In addition, there are now many studies analyzing the chemical composition of the tooth enamel to see whether the creature ate tropical and temperate grasses (C4 plants) or most other types of plants (C3 plants). Wang and Deng (2005) and Martin et al. (2011) did such an analysis of the carbon isotopes, and found that indricotheres consumed a pure C3 plant diet (mainly leaves), just as predicted from their mesowear and other anatomical characteristics.

The vertebrae and ribs of indricotheres are only partially known, with a number of the neck vertebrae, back vertebrae, and most tail vertebrae missing from existing collections (Figs. 2.4, 6.4). Granger and Gregory (1936, pp. 9–13) described some of these from the AMNH Mongolian collection. They noted that even with their small sample, there were a number of different-sized grades from different-sized individuals, as there were for Forster Cooper's (1923a, 1923b) specimens from Dera Bugti. Thus, it was very hard to string together a set of ver-

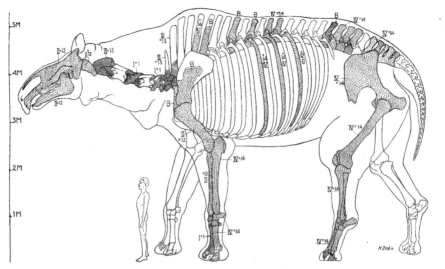

Figure 6.4. Reconstruction of the skeleton of the Mongolian "Baluchitherium grangeri" by Granger and Gregory, as drawn by artist Helen Ziska. Shaded elements are bones that are actually preserved in the Mongolian collections from the American Museum expeditions. Roman numerals I–IV indicate which of the four size classes each bone comes from. (From Granger and Gregory, 1935, 1936.)

tebrae of the same size to make a composite skeleton, since most specimens come from different-sized animals.

Nonetheless, there are some striking features. The first two neck vertebrae, the atlas (which has sockets for the ball joints at the base of the skull) and the axis (the second neck vertebra, which has the rotating joint inserting into the back of the atlas that allows the head to turn) are much wider than they are in most living rhinos, with numerous canals and open spaces that would allow powerful ligaments and thick neck muscles to pass through. This would be expected for an animal with such a huge head that needed strong muscles and ligaments to hold it up. The rest of the neck vertebrae (cervicals) are also very wide, with large flaring ridges of bone that are much wider and more robust than in living rhinos and are penetrated by numerous grooves and canals. These widely flaring flanges (known as zygapophyses in the vertebrae) with lots of room for muscles, tendons, and ligaments, are found not only in indricotheres, but also in the necks of sauropod dinosaurs, which must support the enormous weight of the neck and head with extra-strong muscles and tendons. Granger and Gregory (1936, p. 17) commented "the mid-cervicals are represented in our collection by two vertebrae of colossal size which at first sight look more like the vertebrae of sauropod dinosaurs than like those of even the largest land mammals." Forster Cooper (1923a, p. 38) compared the neck proportions of indricotheres to those of horses, but because the entire sequence of neck vertebrae is unknown, this is hard to ascertain.

A few of the vertebrae (Fig. 6.4) of the upper back (dorsal or thoracic vertebrae) are known, and they show the same tendency to have wide flaring flanges (zygapophyses) on their sides, with lots of holes and canals for the nerves plus the tendons, ligaments, and muscles needed to support the spine. Most of them also bear a long spine that sticks straight up (the neural spine) and would have formed a long "hump" on the back of the animal. This hump, made of high neural spines and their surrounding muscle tissues, is a crucial feature, since it is the place where the nuchal ligament holding up the skull attaches. It is also the attachment point for many of the neck muscles as well.

As can be seen in Fig. 6.4, only a few ribs are known, and many of the vertebrae of the lower back (lumbars) and the entire tail vertebral sequence (caudals) are missing as well. Granger and Gregory (1936, p. 41) compared the ribs that are known to those of the Indian rhino. They pointed out that indricothere ribs were nearly identical to those of modern rhinos except for their much larger size. However, they note that because the legs of indricotheres were so long and their bodies so large, the rib cage would look "decidedly smaller in proportion to the height of the whole animal" compared to the short-limbed living rhinoceroses.

One part of the spinal column that was preserved was the hip region or the sacrum. Although the best-known specimen of the sacrum is badly broken, it seems to be composed of four sacral vertebrae fused together (as they are in most groups of mammals), plus the last vertebra of the thorax (lower back) is fused to the sacrum. This feature is unknown in primitive rhinoceroses, but occurs both in indricotheres and also in advanced rhinocerotids.

If the head and neck and spine and ribcage (the "axial skeleton") of indricotheres is remarkable, even more so are the limbs. As Granger and Gregory (1936, p. 41) noted, they are unusual in many ways. They have become huge and robust in response to the biometric constraints of bearing such a huge weight. In this respect, they are similar to other huge quadrupedal animals, such as elephants and mammoths, or sauropod dinosaurs. These animals show many convergences due to their huge size, so their limbs are known as "graviportal" (Latin for "carrying weight"). But the indricotheres also bear the hallmarks of their long-legged running hyracodont ancestors as well, and in this regard their huge limbs look very different from those of elephants or sauropods.

Graviportal animals have a tendency to lengthen the limb elements nearest the body (the humerus, or upper arm bone, and the femur, or thigh bone), while shortening the lower limb segments (the radius and ulna, or forearm bones, and the tibia, or shin bone). By contrast, animals that are specialized for running have shortened the upper limb elements (humerus, femur) while greatly elongating the lower limb elements (radius-ulna, tibia) and often elongating the hand bones (metacarpals) and foot bones (metatarsals) as well. Indricotheres do not show this degree of specialization for running, but then their ancestors, the "running rhinos" or hyracodonts were not nearly as specialized for running as any of the modern runners, like horses, deer, and antelopes. Nonetheless, they still retain the short

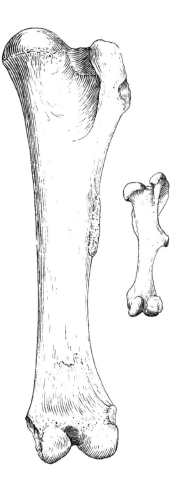

Figure 6.5. Comparison of the femur (thighbone) of Paraceratherium *with that of a contemporary smaller fossil rhinoceros* Epiaceratherium. *(From Granger and Gregory, 1936, fig. 41.)*

upper limb elements and the long hand and foot bones like other running mammals, rather than the long upper limb elements and short hand and foot bones seen in proboscideans and sauropod dinosaurs.

Each individual leg bone is huge, in and of itself, and much more robust and thick than you would expect for smaller animals. Typical of the limb bones is the largest of them, the thighbone or femur (Fig. 6.5). These massive bones typically measure 1.5 m or almost 5 feet in length or longer, only a little shorter that many humans are tall! Only the femora of a few mammoths and elephants, and of some dinosaurs, are longer.

By comparison to normal-sized living and fossil rhinos, the indricothere femur is much thicker and more robust compared to its length, as would be expected with the normal scaling up to gigantic animals. As scientists have noted since the time of Galileo, limbs respond to increased size by developing into thick pillars with relatively robust shafts (compared to their relative length), because even as their cross-sectional area increases as a power of two (i.e., square footage increases), the volume of a giant animal increases as a power of three (i.e., cubic volume increases). Large animals have much higher volume and mass and thus must have more robust limbs relative to those from smaller animals. The other striking feature of the indricothere femur compared to that of more normal-sized rhinos is that the various ridges and flanges that stick out of the shaft (greater, lesser, and third trochanters on the femur) are much more subdued and reduced (Fig. 6.5) in indricotheres. This is probably a consequence of scaling; as the shaft of the limb gets larger and thicker, it tends to diminish the importance of these ridges. A similar reduction is seen in the limb bones of other huge animals, such as elephants and sauropod dinosaurs. Also, the columnar posture of the limbs (compared to the partially bent posture of the limbs of most smaller animals) means there is less need for large muscles to hold the legs in a bent, flexed position.

The final feature of the indricothere skeleton is the most remarkable of all: the lower limbs, and especially the hands and feet. In most graviportal animals such as elephants and sauropods (Fig. 6.6), there is a tendency to greatly shorten and fuse together the lower limb elements. In elephants, the radius and ulna are short and fused into one forearm bone, while the shin bone, or tibia, is very short

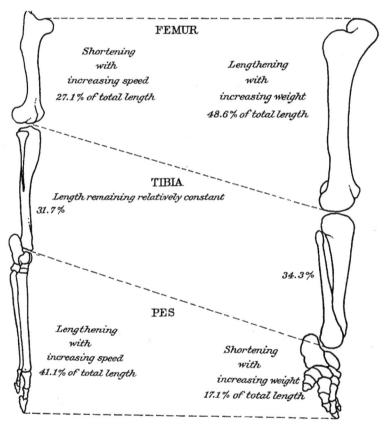

Figure 6.6. Comparison of limb bone proportions between a horse, a running mammal (left), which shortens the upper limb segment (humerus or femur) and lengthens its wrist and ankle and toe bones, versus a graviportal mammal like an elephant, which has long robust upper limb segments and short compressed hand and foot bones. (After Osborn, 1929, Fig. 670.)

and robust. However, in indricotheres, there is no fusion of the radius and ulna, and both the radius-ulna and the tibia are relatively long compared to those in elephants or sauropods.

Both elephants and sauropod dinosaurs respond to their greatly increased weights by developing highly compressed hand and foot bones; in some cases, these bones are squashed into pancakes compared to the hand and foot bones of normal mammals. In animals this heavy, the extra leverage of the hand and foot bones and long fingers and toes (so important in a horse or deer) is no longer of great importance, and the hand and foot become a mashed-together collection of flattened bones that support a thick pad on which they walk.

Contrast this with the hands and feet of indricotheres (Fig. 6.7). First of all, the individual hand and foot bones (metacarpals and metatarsals) are huge (Fig. 6.7A), some almost 50 cm (0.5 m, or 20 inches long), as would be expected for an animal this large. The surprise, however, is that the metacarpals and metatarsals (together, they are known metapodials) are not squashed into short, stubby, fat bones as they are in elephants and sauropods (Fig. 6.6). Instead, they are long and relatively thin. Only the finger and toe bones (phalanges) are squashed into flat disks, as is typical of most graviportal animals (Fig. 6.7B). This peculiar com-

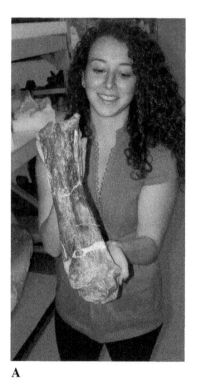 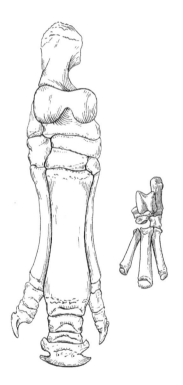

A   **B**

Figure 6.7.A. An individual foot bone (metapodial) from Paraceratherium, *showing how immense it is compared to an average person and also how long and slender it is. (Photo by the author). B. The foot bones of* Paraceratherium *compared to a contemporary rhinoceros* Epiaceratherium. *(From Granger and Gregory, 1936, Fig. 42.)*

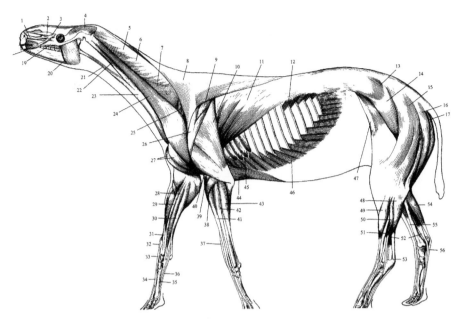

Figure 6.8. Reconstruction of the musculature of the smaller indricothere Juxia. *(After Qiu and Wang, 2007.)*

bination of long metapodials and short finger and toe bones can only be explained as a relict of the hyracodont ancestry of the indricotheres. The hyracodonts, or running rhinos, are characterized by their unusually long metapodials compared to other early rhinos. The indricotheres, even though they were far too large to be good runners (and probably had no real predators that they needed to run from), still retain this hallmark of their running ancestry, in spite of all the evolutionary forces that were molding their limbs to the elephantine "graviportal" type.

Putting Flesh on the Skeletons

Reconstructing extinct fossils as living animals is not as simple and straightforward as it seems. Many artists just take the sketch of the mounted skeleton and arbitrarily put some muscles and skin on it (Fig. 6.8), although there are some artists (like the legendary Jay Matternes, Carl Buell, and Mauricio Anton) who do a better, more rigorous job of restoration based on their detailed training in anatomy.

A good restoration starts with the skeleton and its mount. Osborn himself (1923b) admitted that his first restoration of the animal (Fig. 6.9) was rushed and inaccurate because it was based on only part of the material that returned from Mongolia after the 1922 expedition. This version attempted to force the bones into the proportions of modern rhinoceros, resulting in legs that were too short

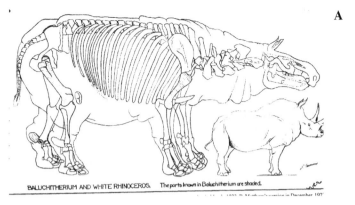

BALUCHITHERIUM AND WHITE RHINOCEROS. The parts known in Baluchitherium are shaded.

A

B

Figure 6.9. A. Osborn's original restoration of "Baluchitherium" from Mongolia, before most of the Mongolian specimens had been prepared or studied. It was too much like a modern rhino, with an incorrectly short neck and legs (From Osborn, 1923b). B. A later restoration with longer neck and legs, based on more complete material found since 1926 and Granger and Gregory's (1936) reconstruction. For many years, it was on display as a life-sized bas-relief plaster sculpture on the wall of the fossil mammal hall at the American Museum. The people in front include Walter Granger (second from right) and William King Gregory (extreme left). (Image number 410739; courtesy American Museum of Natural History Library.)

and flexed (especially the hind legs), and a neck that was too short and thick. Osborn (1923b) contrasted that with his more current restoration, which is similar to the Helen Ziska illustration in Granger and Gregory (1935, 1936) (Fig. 6.3). In this version, the neck and legs are much longer and more slender and the back is straighter. Since the Ziska drawing in Granger and Gregory (1935, 1936), other artists have taken liberties with the original data, so it common today to find images of indricotheres that are even more delicate and gracile, with long slender legs and neck, and no evidence that the artists knew much about the original skeleton.

What most people do not realize unless they read the original Granger and Gregory (1935, 1936) paper is that the composite skeleton is based on a range of specimens from different individuals of different sizes, so it is a chimera that may not really represent the true proportions of any one individual indricothere. Fortelius and Kappleman (1993) discussed this at length and pointed out its implications for weight estimates (see final section).

A very different looking reconstruction was drawn by N. Yanshinova, under the direction of Konstatin Flerov, and published by Gromova (1959) (Fig. 6.10). It was based on the mounted partial skeleton from the Aral Sea region (Fig. 2.4), which lacked a skull (they used a cast of the American Museum Mongolian skull, Fig. 6.1, as a replacement). This skeleton reconstruction is much more lightly built than the Ziska-Granger-Gregory skeleton, although both skeletons lack most of the neck vertebrae needed to get the true length and proportions of the neck.

From these artistic reconstructions, a number of three-dimensional replicas have been produced. The most famous is the fiberglass model that was once on display in Morrill Hall at the University of Nebraska State Museum in Lincoln, Nebraska (see Frontispiece). This model was eventually removed and sold to the Wyobraska Wildlife Museum in Gering, Nebraska (www.wyobraskawildlifemuseum.com), where it still stands in the museum's main facility, a former high-ceilinged railroad depot building. In 1995, I was filmed in an episode, "Are Rhinos Dinos?" for the BBC documentary series *Paleoworld* using this specimen. It is quite an experience being able to walk completely under the belly of this model without stooping (and I'm almost 6 feet tall).

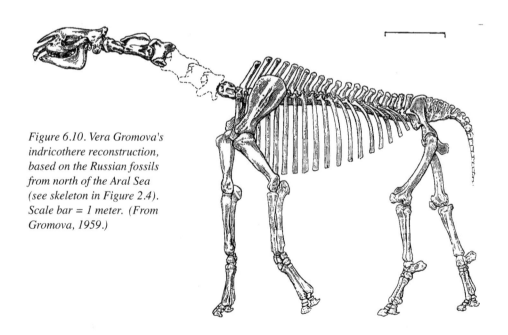

Figure 6.10. Vera Gromova's indricothere reconstruction, based on the Russian fossils from north of the Aral Sea (see skeleton in Figure 2.4). Scale bar = 1 meter. (From Gromova, 1959.)

Besides the issue of proportions and neck length, there are other issues that artists and scientists who attempt to reconstruct extinct beasts must resolve. In the case of indricotheres, we have only the bones, and only partial skeletons at that. There is no direct information about the color of the animal, its skin texture, what it ate, how it walked, or how it behaved or sounded. All of this information, so often a crucial part of the CG animations that now dominate most documentaries about prehistoric life, is entirely conjectural and cannot be determined directly from the bones. The usual approach is to model indricotheres on the basis of living rhinos, with thick, gray, hairless skin with numerous folds, although we have no skin impressions or mummified specimens to test this idea one way or another. The behavior and colors and sounds of the animals in these CG animations (such as in the documentary "Walking with Prehistoric Beasts") are completely imaginary and have no basis in any real-world data. Although most scientists are aware of this, a surprising number of people who watch these TV shows are stunned that so much of the show is pure fiction for entertainment, rather than science. The only real science in these shows is the interviews with expert paleontologists and the pictures of bones and fossil localities.

Constraints on Giants

Although most of the stuff you see in CG animations of prehistoric beasts (including the indricotheres) is mostly guesswork, there are living analogues that can give us some guidance about indricothere biology. The best models might be elephants, which approach indricotheres in body size (see the next section). There are certain constraints about life with such a large body size for elephants that must also apply to the indricotheres:

Thermoregulation: Elephants have a huge body volume and mass compared to their surface area (remember, volume increases as a cube while area only increases as a square). As the debate about hot-blooded dinosaurs back in the 1980s revealed, such huge animals with an endothermic physiology (that is, they generate their own body heat from metabolism) have a severe problem getting rid of excess body heat, especially if they live in warm climates (Sander et al., 2011). The living elephants have huge ears like radiators to shed excess body heat from their bloodstream, and it is reasonable to infer that indricotheres did too. African elephants and rhinos and hippos spend much of their daytime resting in the shade or wallowing in waterholes and mud puddles to cool down, and the indricotheres must have too. Elephants, rhinos, and hippos feed and move mainly at night, as indricotheres must have done. Elephants and rhinos both have largely naked skin since hair holds in body heat, which is why such elephant-like naked gray skin seems appropriate for indricotheres. Large-bodied endothermic mammals are in a constant battle to dump body heat and avoid overheating.

Digestion: There are certain other constraints for large-bodied herbivores as well. All herbivores eat large amounts of cellulose in their diets, which is a relatively indigestible carbohydrate. Most plant eaters must use some kind of specialized gut bacterium in their digestive tract to break down the cellulose and release the nutrients. Such a breakdown requires fermentation and takes time to absorb the nutrients from the fermentation process into the lining of the intestines. As Janis (1976) and Mellett (1982) pointed out, there are two basic types of herbivore digestion: foregut fermenters and hindgut fermenters.

The only living foregut fermenters are the ruminant artiodactyls (camels, cattle, sheep, goats, antelopes, deer, and pronghorns), which do this by "ruminating" using a four-chambered stomach. The first chamber, the rumen, is a digestive vat full of bacteria, so that when they swallow a bite of partially chewed plant material, it goes immediately into the rumen where it begins bacterial breakdown. Later, when they are resting, ruminants regurgitate some of the contents of their rumen back into their mouths, where they can "chew their cud" and break the material down further before swallowing it again. By the time the food reaches the lining of their intestines, it is highly broken down into nutrients and easily absorbed. Thus, ruminants use nearly every bit of their food efficiently and can survive on relatively small amounts of good-quality vegetation. But if they eat too much high-quality vegetation, they can become bloated, and their rumen can swell and even rupture and kill them with all the gas released from the rapid bacterial fermentation.

The rest of the herbivorous mammals are hindgut fermenters. These include the perissodactyls (odd-toed hoofed mammals, today including the horses, tapirs, and rhinos), the elephants, the non-ruminant artiodactyls (pigs, peccaries, and hippos), and other herbivores such as rabbits and some primates. Instead of a highly specialized foregut with a rumen, they have the normal mammalian digestive tract, with an esophagus, acid-filled stomach, and finally intestines for absorption. Most have a pouch off the intestine called a caecum that is the primary location of bacterial fermentation. Lacking a rumen, the hindgut fermenters pass the mostly undigested cellulose through the digestive tract until it reaches the caecum, but bacterial fermentation only just starts in the caecum before the food goes through the remaining intestines and is then excreted.

Consequently, hindgut fermenters get relatively little nutrition out of each bite of fodder and must eat much larger volumes of mostly low-quality food (especially grasses) to get enough nutrients to live on. Most hindgut fermenters, like horses, rhinos, and elephants, are by necessity high-volume, low-efficiency eaters, and eat huge volumes of material just to survive, since they are so poorly adapted at extracting nutrients. When you see the feces of these animals (like the "road apples" of horses), they are typically full of undigested plant matter compared to the "cowpies" of a ruminant, or the tiny pellets of a deer or pronghorn.

Rabbits are a special case. If you have ever kept rabbits in a hutch, you will notice that they eat their own feces. This gives them a chance to run the food through their gut a second time after the bacterial fermentation has had time to work, and to get more nutrition this way.

For these reasons, there are certain things we can say with confidence about indricothere feeding dynamics. Because they were not ruminant artiodactyls, they had to be hindgut fermenters, like horses, other rhinos, and elephants, so they must have consumed and processed huge amounts of food in a day, just as elephants do now. Their feces would have been full of undigested plant material, just like those of a horse or an elephant. Like almost all large herbivores, they must have had a big part of their abdomen occupied by their large digestive tract, giving them a large bulging "gut" like that of an elephant. The fermentation in their gut, by the way, creates additional body heat, which exacerbates the problem they have of producing excess body heat to begin with.

Locomotion and Home Range: As an animal increases in body size, the stresses on their limb bones increase even more because of the power of three expansion of volume and the corresponding mass increase. Models of the dynamics of large dinosaurs show that they could not have run very fast or their limbs would have broken (Alexander, 1989). Modern elephants also cannot run very fast compared to true specialized runners like antelopes, horses, or cheetahs. Their maximum speed in an all-out charge clocked at only 29 kph (18 mph), but their normal walking speed is about 10–19 kph (6–12 mph). Remember, they have an advantage in their speed because they have much longer limbs and strides than any other animal. Given that indricotheres were just slightly larger than modern elephants, we can predict that they too would have not been fast runners, but ambled along at a moderate pace like that of an elephant.

However, African elephants are capable of moving enormous distances (typically 32 km or 20 miles) in the course of a day, migrating from one food source to another. To support their food needs of about 140 kg (300 pounds) of food each day, elephants need huge home ranges of 750–1500 square km (300–600 square miles). Consequently, huge home ranges and long migrations would be expected of indricotheres as well (as discussed in Chapter 5), especially if they lived in a harsh desert scrub setting with scarce food sources that were easily wiped out (see Chapter 7). A similar model has been proposed for the large sauropod dinosaurs, which lived in a scrubby, semi-arid habitat in the Late Jurassic time (Morrison Formation) and probably roamed in small herds from one patch of trees to another (Farlow et al., 2010).

Predators and Life Habits: Certain other ecological parameters are also dictated by the giant body sizes of elephants and indricotheres. Once they reach a large enough body size, healthy elephants have no natural predators—not even lions or tigers are foolish enough to tackle them. (This has all changed now with human poaching, which has nearly wiped out elephants in the wild.) Only the babies and young calves are vulnerable to predators, and in elephant herds, there is a strong matriarchal hierarchy so that every calf is closely protected not only by its mother, but also by its sisters, grandmothers, aunts, great-aunts, and other close female relatives. Almost all natural deaths of elephants occur when predators catch vulnerable calves. Almost a quarter of the calves born to Asian elephants are lost to tigers before they reach their first birthday. If indricotheres maintained small herds in the elephant mode, such freedom from predation except for the young would also be true. However, in the Bugti beds there are gigantic croco-diles (*Crocodylus bugtiensis*) that are 10–11 m (33–36 feet) long! These would have been large enough to attack almost any indricothere that might be at the edge of the river to drink. Indeed, many of the specimens from the Bugti beds have crocodile tooth marks on them (P.-O. Antoine, pers. comm., 2011).

There is also a well-known relationship between the gestation period, size of the litter, and body size. Elephants have the longest gestation period of any land creature (22–24 months, or about two years). The females do not reach sexual maturity until they are ten years old, and may produce a single calf every three to four years, the slowest reproductive rate of any mammal. Such could be ex-pected of indricotheres as well, since the constraints of growing to such large body sizes and having such large calves are very similar to those on elephant re-production. Like elephants, indricotheres would be expected to grow quickly at first and then grow relatively slowly once they reached maturity.

Physiologists have shown that there is strong relationship between body size, metabolic rate, and blood pressure. An elephant has a relatively slow metabolic rate. Its heart beats only 30 times per minute, while humans have a pulse of 60 beats per minute, and hamsters have a pulse rate of over 450 beats per minute! The indricothere heart would have had a pulse rate close to that of an elephant, but probably a bit higher. This is because it must have also been able to exert a blood pressure close to the 300 mm Hg that giraffes produce (humans typically have a BP of 120) to be able to lift its head so far above the ground without faint-ing.

Finally, there is also a well-known scaling of longevity with body size, with larger animals (and their slower heart rates) living longer. A rodent typically lives no longer than 3–5 years, a cat or a goat about 15 years, a pig or monkey about 20–25 years, and a cow or giraffe about 25–30 years. Elephants typically live 35–50 years in the wild (at least they did until recent years, when poaching has nearly wiped them out), and the record is 71 years. Similar life spans could be expected of indricotheres as well.

Weight Problems

As the largest land mammal that has ever lived, there is much fascination with maximum body size and weight estimates for extinct indricotheres. Most often you see sources parroting each other around a common estimate of 20 metric tonnes (20,000 kg), although Alexander (1989) estimated 34 tonnes and Savage and Long (1986) quoted 30 tonnes. Prothero and Schoch (2002) followed Alexander (1989) and used the number of 34 tonnes, and there have been a wide range of estimates all over this range. As Fortelius and Kappelman (1993) showed, these estimates came from some questionable assumptions and extrapolations by Osborn, Granger, and Gregory based on their partial skeleton (Fig. 6.3). That skeletal reconstruction is a composite, a chimera based on bones from individuals of at least four different size classes, so it is not a very reliable source for the body size of an actual animal.

Most of the weight estimates are based on the combined head and body length (HBL) of the animal, which may be exaggerated if Granger and Gregory's (1935, 1936) reconstruction drawn by Helen Ziska is inaccurate. As Fortelius and Kappelman (1993, Table 1) showed, different authors estimated a wide range of HBL values, from Gromova's estimate of 740 cm to Granger and Gregory's (1936) maximum of 870 cm, which drops to 621 cm if you use just the smallest individuals. Using the weight equations of Damuth (1990) for converting HBL to mass, this gives weight estimates from as low as 8.4 tonnes (for the smallest specimens) to 24 tonnes for the largest specimens.

Fortelius and Kappelman (1993, Table 2) looked at a spectrum of other methods to estimate body size, based on measurements of the skull. The measurements of skull dimensions gave a range of masses form 7–16 tonnes, with a single high estimate of 19.8 tonnes, which is probably an overestimate based on the creatures from which this estimate comes. Using the teeth as an estimator gives values in the range of 5–15 tonnes, although indricothere teeth seem unusually small compared to the size of their skulls and bodies. Estimates based on upper limb bones (humeri and femora) gave a range of 5 to 15 tonnes. Overall, their estimates placed most specimens of indricotheres in the 9–15 tonne range, with only a few specimens giving values as high as 20–24 tonnes.

Figure 6.11. The giant deinotheres were primitive proboscideans with downturned lower tusks and were the largest land mammals in most of Eurasia and Africa during the Miocene, replacing the indricotheres. (After Scheele, 1955.)

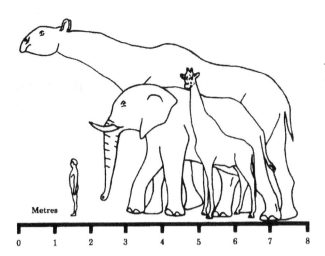

Metres

0 1 2 3 4 5 6 7 8

Gingerich (1990) developed his own method of estimating body mass based on various limb bone measurements. He obtained weight estimates of about 9 tonnes for smaller indricotheres, and about 14–15 tonnes for the largest specimens. As he pointed out, however, there are serious problems with any of these estimates. Not only is it difficult to get accurate predictions based on just a few bones from four different size classes, but indricotheres also had very differently proportioned limbs compared to any other living mammal.

Once we drop the outrageous overestimates of the 30 tonne range and use a more realistic 15–20 tonne estimate for the biggest indricotheres, the huge difference in size between indricotheres and elephants vanishes. Rather than "weighing as much as five elephants" as is often quoted, indricotheres were typically in the size range of the largest elephants and mammoths. The largest living elephant specimen ever measured was a bull elephant killed in Angola in 1955 and now on display in the Smithsonian, which weighed 10 tonnes, but most elephants weigh about 5–6 tonnes (Prothero and Schoch, 2002, pp. 182–183). However, Fortelius and Kappelman (1993) pointed to examples of limbs of the mammoth *Mammuthus trogontherii* and the *Deinotherium* (with its downward-flexed tusks—Fig. 6.11) that suggest weights of 13 tonnes to possibly as high as 20 tonnes (Christiansen, 2004). Saarinen (pers. comm., 2011) looked at limb bones of *Deinotherium* from European museums and estimated a range of weights of 7.4–17.4 tonnes, with a mean of 11.2 tonnes. Most fossil proboscideans give weight ranges between 3–10 tonnes, which would make them somewhat lighter than indricotheres. Thus, we should be cautious about calling indricotheres the "largest land mammals that ever lived." They were certainly taller than modern elephants or other living mammals (Fig. 6.12), but their mass was roughly equivalent to the largest mammoths and deinotheres.

These weight estimates are not just a piece of trivia to pop into a textbook, but have implications as well. A number of ecologists and physiologists have speculated about the maximum body size that mammals can obtain, based on constraints due to metabolic factors. Economos (1981) was the first to do this, using mathematical estimates of the metabolic costs of gravity. Based on this method, he concluded that a land mammal cannot theoretically reach sizes of much greater than 20 tonnes, which seems to correspond to our upper limit for largest indricotheres and mammoths. Clauss et al. (2003) looked at the nutritional constraints of large body size, particularly in contrasting the relatively efficient foregut-fermenting ruminants (which tend not to grow to huge sizes) versus the inefficient hindgut fermenters like elephants, rhinos, and hippos. Based on these constraints, they found that Fortelius and Kappelman's (1993) estimate of 11–15 tonnes for indricotheres was more consistent with digestive constraints than the higher estimates of 20 tonnes or greater.

Thus, we must be careful when quoting old numbers from early authors about the weights of extinct creatures. Indricotheres probably weighed only in the 10–15 tonne range and maxed out at 20 tonnes on the largest individuals. It is very unlikely that there were any in the 30–35 tonne range, as is so often cited.

7

Paradise Lost

Greenhouse of the Dinosaurs

Fifty-five million years ago (the earliest Eocene Epoch), the planet was in its warmest phase since the "greenhouse of the dinosaurs" during the latter part of the Mesozoic (Prothero, 2009). There were crocodiles in the polar regions, along with a wide variety of mammals (including the earliest rhinocerotoid *Hyrachyus*). The fossil plants found above the Arctic and Antarctic Circles look nothing like the Arctic tundra or Antarctic ice caps that are found there today, even though they must have experienced six months of darkness. Instead, during the Eocene Epoch there were broad-leaved evergreens, including palm trees and cycads, above 61° north latitude in Alaska, indicating average temperatures around 18°C (65°F). There were broad-leaved deciduous forests and even rich coal beds, indicating dense forest vegetation. Spitzbergen produced a flora that could not have tolerated freezing. Ellesmere Island in the Canadian Arctic, which today lies at 78° north latitude and was not far from that latitude in the early Eocene, produced similar fossil plants. It also yields fossil alligators, pond turtles, land tortoises, and monitor lizards, as well as garfish and bowfin fish. Most of these animals are typical of subtropical climates today, and none can tolerate freezing for long. Alligators are limited by a mean coldest month temperature of 10°C. There is also a surprisingly diverse fauna of mammals typical of the late early Eocene in North America, including tapirs, horses, brontotheres, primates, rodents, colugos, and rhinos.

In the high latitudes of the Southern Hemisphere, the story was the same. In Australia there was a subtropical-temperate forest dominated by typical Southern Hemisphere conifers, such as *Araucaria,* which included the Norfolk Island pine and "monkey puzzle tree," and podocarps, along with flowering plants such as the Proteaceae (including the familiar flowering shrub *Banksia*) and laurels. In the swampy areas, *Nypa* palms and tree ferns were common and coals formed. According to Greenwood et al. (2003), the mean annual temperature for this region was 16–22°C (up to 72°F), with the coldest mean temperature no lower than 10°C (50°F), and a mean annual precipitation of more than 150 cm per year.

How could the polar regions be so balmy, even though they experienced six months of darkness? When the evidence from Ellesmere Island and other polar localities was first discovered in the 1980s, paleoclimatologists struggled for an answer to this. Paleomagnetic pole positions for Eocene rocks of the high Arctic show that they have not drifted significantly northward since they formed; so we can rule that possibility out (McKenna, 1980, 1983). Ellesmere Island, for example, was at 75° north latitude in the Eocene, not significantly different from its present latitude. Creber and Chaloner (1984) re-examined the botanical evidence cited by Wolfe (1980) and concluded that even with six months of darkness, the polar regions received sufficient sunlight for a seasonally productive forest. The main limiting factor was temperature. As long as it did not get too cold, the forests could grow during the "midnight sun" summers and remain dormant during the six months of darkness. The modern dominance of tundra vegetation in the Arctic is dictated by the cold, dry conditions, not by the six months of darkness. As I discussed in my book *Greenhouse of the Dinosaurs* (2009), all the evidence now shows that the carbon dioxide levels of the early Eocene were nearly as high as those of the Cretaceous (perhaps 1000–2000 ppm, compared to about 300 ppm today). It was a true "greenhouse planet."

Similar warm and wet subtropical to temperate conditions could be found in regions that are now buried under snow each winter. Fossil plants in Wyoming, North Dakota, and Montana demonstrate that conditions warmed to mean annual temperatures as high as 21°C (70°F), and even the mean annual cold month temperature could be no lower than 13°C (55°F), because most of the plants are intolerant of freezing. These plants also suggest that the climate was very wet, with mean annual rainfall in excess of 150 cm (60 inches). By contrast, western North Dakota today has a steppe climate. The mean annual temperature is only 5°C (41°F), and the spread between daily extremes ranges over 33°C (over 90°F). In North Dakota or eastern Montana, it is not at all unusual for the temperatures on a hot spring or fall day to start out above 32°C (90°F), then drop below freezing in a matter of hours as an Arctic cold front moves in.

From the evidence of floras in the Bighorn Basin of Wyoming or the Williston Basin of Montana and North Dakota, we can visualize a dense tropical forest much like that found in modern Panama. Tall trees formed a dense canopy, with vines and lianas growing all around them. The fossil plants include many the tropical groups, including citrus, avocado, cashews, and paw paw trees. Many of these plant genera are found today only in the jungles of southeast Asia or tropical Central America. In addition to the direct evidence of the plant fossils, there is information in the striking color bands that stripe the badlands slopes. Each band represents an ancient soil horizon, and in many places there are hundreds of them stacked on top of each other, representing millions of years of the early Eocene. Each represents another episode of floodplain mud deposition, followed by the development of plants and a soil horizon, and then another episode of flooding, which buried the old soil. According to Tom Bown and Mary Kraus (1981, 1987), these ancient soils were deposited on broad floodplains bordering meandering rivers, much like those of the modern Amazon.

The same pattern can be seen in other temperate and tropical regions throughout the early Eocene. Even the Pacific Northwest and southern Alaska were relatively warm (25°C) and wet, blanketed with broad-leaved evergreen forests, with abundant vines and lianas, and many plants of tropical Asian affinities.

From the London Clay found in the basements of London comes an important early Eocene flora. As we saw in Montana, there are mostly tropical trees and shrubs, lianas, including cinnamon, figs, magnolias, palms, laurels, citrus, paw paw, cashews, laurels, and vines such as moonseed, icacina, and grapes. Collinson (1983) and Collinson and Hooker (1987) showed that 92 percent of these plants have living relatives in the jungles of southeast Asia. Fringing the coasts of the tropical jungles were mangrove swamps full of *Nypa* palms, also restricted to southeast Asia today. From this evidence, the average temperatures in London were about 25°C (77°F) compared to the modern average of 10°C (50°F). Instead of the cold, foggy London of Sherlock Holmes, London was as warm and tropical as Singapore.

Anywhere we find fossil floras of early Eocene age in temperate or tropical regions around the world, we encounter a similar story. Floras from China (Guo, 1985), Siberia (Budantsev, 1992), India (Mehrotra, 2003), and southern South America (Romero, 1986) all show the same tropical-subtropical patterns, even though many of these regions were at fairly high latitudes and inland locations. Naturally, the few floras known from tropical regions, such as Panama, show that conditions were hot and wet there in the early Eocene, as they are today (Graham, 1999).

The Big Chill

Contrast these conditions in the early Eocene with those of the early Oligocene (about 33 Ma), the world of the indricotheres in Asia. The White River Group rocks of the Big Badlands of South Dakota preserve quite a bit of detail about the history of many kinds of organisms. Retallack (1983) studied the color bands visible in the Badlands sections and found that they were paleosols, or ancient soil horizons. Those from the upper Eocene Chadron Formation were formed under forests with closed canopies of large trees (the huge root casts are particularly conspicuous) with between 500–900 mm (20–35 inches) of rainfall per year. In the overlying lower Oligocene (Orellan) Brule Formation, the paleosols indicate more open, dry woodland with only 500 mm (20 inches) of rainfall per year. Emmett Evanoff studied the sediments of eastern Wyoming (Evanoff et al., 1992) and found that the moist Chadronian floodplain deposits abruptly shifted to drier, wind-blown deposits by the Orellan. In the same beds are climate-sensitive land snails. According to Evanoff et al. (1992), Chadronian land snails are large-shelled taxa similar to those found in wet subtropical regions, like modern Central America. Based on modern analogues, these snail fossil indicate a mean annual temperature of 16.5°C (63°F) and a mean annual precipitation of about 450 mm (18 inches), very similar to the results obtained by Retallack (1983) for neighboring South Dakota. By contrast, Orellan land snails are drought-tolerant, small-shelled taxa indicative of warm-temperate open woodlands with a pronounced dry season. Their living analogues are found today in Baja, California.

The amphibians and reptiles suggest similar trends of cooling and drying in the early Oligocene (Hutchison, 1982, 1992). The Eocene is dominated by aquatic species (especially salamanders, pond turtles, and crocodilians) that had been steadily declining in the middle and late Eocene. Crocodiles were gone by the Chadronian, but there are a few fossil alligators that have been recovered from the Chadron Formation. By the Oligocene, only land tortoises are common, indicating a pronounced drying trend. In fact, these tortoises (*Stylemys nebraskensis*) are so common in the Orellan that these beds were originally called the "turtle-oreodon beds" after their two most common vertebrate fossils.

Land plants are not well-preserved in the highly oxidized beds of the Big Badlands (except for the durable hackberry seeds, which are calcified while they are alive), so we must look to other regions to understand the floral change. The rest of North American floras show a clear trend. Based on leaf-margin analysis, Wolfe (1971, 1978, 1985, 1992) suggested that mean annual temperatures in North America cooled about 8–12°C (13–23°F) in less than a million years. This is by far the most dramatic cooling event of the entire North American floral record and was the original basis for the phrase "Terminal Eocene Event" (even though revised dating now places it in the early Oligocene). Perhaps Wolfe's (1971) earlier phrase "Oligocene deterioration" would be a better term.

The problem with the earlier analyses published by Wolfe (1971, 1978) was due to confusion about dating. Most of these geochronological problems have now been cleared up (Wolfe, 1992; Myers, 2003), but further refinement of dating is always valuable to test previous hypotheses. The Rocky Mountains of central Colorado yield several important floras that span the Eocene-Oligocene transition. The floras of the famous late Eocene Florissant Formation (Evanoff et al., 2001; Prothero and Sanchez, 2004) are dated at 34.07 ± 0.10 Ma. This Florissant flora records the final phase of late Eocene warmth in the North American floral climatic curve of Wolfe (1978) before the early Oligocene deterioration. Even though it was at 2000–3000 m elevation in the Eocene, the Florissant flora is believed to represent warm-temperate climatic conditions of moderate rainfall and a mean annual temperature of 13–14°C (Meyer, 2003), compared to modern mean annual temperatures of 4°C. Slightly younger than Florissant is the late Eocene Antero flora. Durden (1966) reported a flora that was not too different from that of Florissant. As Prothero (2008) demonstrated, there is then a gradual cooling in the high-altitude Rocky Mountain floras, from the early Oligocene Pitch-Pinnacle flora to the late Oligocene Creede flora.

The longest and most complete sequence of floras spanning the Eocene-Oligocene transition occurs in the Eugene and Fisher formations near Eugene, Oregon. Recent high-resolution lithostratigraphy, biostratigraphy, tephrostratigraphy, ^{40}Ar/^{39}Ar dating, and magnetostratigraphy by Retallack et al. (2004) allow fine-scale dating and correlation of the classic floras and marine invertebrate faunas in this sequence. The paratropical middle Eocene Comstock flora is dated at 39.7 Ma, and the flora suggests a mean annual temperature of 22.4°C with a warm-month mean of 26.5°C and a cold-month mean of 7.0°C and a mean annual range of temperature of 19.5°C, all suggestive of warm subtropical conditions. The late Eocene Goshen flora was one of the original bases for Wolfe's (1978) concept of a late Eocene warming event. It is now dated at 33.4 Ma and yields a mean annual temperature of 19.7°C with a warm-month mean of 25.1°C and a cold-month mean of 6.8°C and a mean annual range of temperature of 18.3°C, not significantly different from the warm subtropical conditions of the Comstock flora. The Rujada flora, dated at 31.3 Ma, shows temperature estimates consistent with that of a post-deterioration flora, with a mean annual temperature of 13.0°C, a cold-month mean temperature of 2.4°C, a warm-month mean temperature of 23.6°C, and a mean annual range of temperature of 21.2°C (Retallack et al., 2004). Stratigraphically above the Rujada flora are the early Oligocene Coburg and Willamette floras. They are dated at 30.9 Ma for the Coburg flora, and 30.1 Ma for Willamette flora. The Willamette flora yields a mean annual temperature of 13.2°C with a warm-month mean of 20.8°C and a cold-month mean of 6.2°C, and a mean annual range of temperature of 14.6°C, and the much less well-known Coburg flora is very similar. These floras clearly suggest that the Oligocene deterioration had taken place by 31.3 Ma. Thus, the Eugene-Fisher floral sequence places the climatic change in the earliest Oligocene (consistent with the global record), somewhere between 33.4 and 31.3 Ma.

In the John Day region of central Oregon, the Clarno Formation yields floras relevant to this discussion. The Iron Mountain assemblage of the Bridge Creek flora (Meyer and Manchester, 1997; Myers, 2003) is ^{40}Ar/^{39}Ar dated at 33.62 Ma (at the Eocene-Oligocene boundary) and yields a post-deterioration mean annual temperature of 10.3°C, with comparable results for winter mean annual temperature of 0.9°C and summer mean annual temperature of 17.7°C (Myers, 2003). Just above this flora is the Fossil High School assemblage of the Bridge Creek Flora, which is dated at 32.6 Ma, and yields a mean annual temperature of 12.1°C with a warm-month mean of 19.0°C and a cold-month mean of 3.1°C, also suggesting a post-deterioration flora. Thus, the climatic event in central Oregon seems to happen before 33.6 Ma (a date that is right at the Eocene-Oligocene boundary), not within the early Oligocene.

In summary, the cooling event found during the earliest Oligocene in the global climatic record seems to cool gradually over the late Eocene and Oligocene

(Prothero, 2008), rather than abruptly as Wolfe (1978) originally thought.

Despite these changes in the soils, land plants, land snails, and reptiles and amphibians, the change in the mammalian fauna is not that impressive (Prothero, 1994; Prothero and Heaton, 1996; Prothero, 1999). Most of the archaic Eocene taxa (especially the forest dwellers and arboreal forms) were already gone by the last Eocene. A few groups, such as the brontotheres, the camel-like oromerycids, the mole-like epoicotheres, and two groups of rodents, did die out near the end of the Chadronian, but none were around to witness the big early Oligocene climatic deterioration. Most of the mammals that were present before the climatic crash showed no change whatsoever, except for a dwarfing event in one lineage of the oreodont *Miniochoerus*. Apparently, the groups that were present in the late Chadronian were already adapted to the drier, more open woodlands habitats, so the vegetational change did not make that much difference—or else the responsiveness of mammals to short-term changes in climate has been oversold, and they are not as sensitive as we have long believed (Prothero and Heaton, 1996; Prothero, 1999).

The mammals that lived in what is now the Big Badlands during the Orellan were very much like those of the Chadronian, minus the huge brontotheres. Oreodonts (a group of extinct artiodactyls distantly related to camels) were by far the most common creatures, with hundreds of their skulls discovered in almost every meter of the section. Early gazelle-like camels (*Poebrotherium* and *Paratylopus*) without humps were also common, as were the huge pig-like entelodont *Archaeotherium* and the tiny deer-like artiodactyls without antlers, *Leptomeryx* and *Hypisodus*. Among perissodactyls, the primitive three-toed horses *Mesohippus* and *Miohippus* were common, as were the hippo-like *Metamynodon* rhinos (Fig. 4.6), the long-legged Great Dane-size running rhino *Hyracodon* (Fig. 4.8), and the true rhinoceros *Subhyracodon*. Tapirs, on the other hand, were very rare after their great diversity in the middle Eocene. Small mammal faunas were dominated by rodents (especially modern groups like hamsters, pocket gophers, beavers, and squirrels), plus abundant rabbits, and a whole suite of insectivorous mammals. Preying upon the herbivores was the last of the archaic creodonts, *Hyaenodon* (Fig. 3.6C), plus a much greater diversity of true carnivorans: the early dog *Hesperocyon* (which looked more like a weasel), the first members of the weasel family, plus primitive amphicyonids (known as "beardogs," they are an unique extinct family) and abundant nimravids, which converged on cats and sabertooths, even though they are not closely related to cats at all.

Most of the herbivorous mammals had fairly primitive, low-crowned dentitions for eating leaves and shrubs, and there is no evidence of grasslands or animals with high-crowned teeth for eating them yet. However, animals with extremely primitive low-crowned teeth (such as brontotheres) were gone by this time, so that whatever plants they fed on, the softest vegetation must have disappeared. The absence of tree-dwelling mammals like primates and multituberculates also shows that the dense forests must have vanished, since almost all the small mammals (even the squirrels) appear to be adapted for life on the ground. This "White River Chronofauna" was stable and well established, and would remain relatively unchanged from the late Eocene (Chadronian) until well into the Miocene.

La Grande Coupure

In Europe, there are no great desert badlands exposures rich with fossils like those in North America or Mongolia. Instead, the best fossil record of terrestrial life in the Eocene and Oligocene comes from pockets and caverns in the limestone bedrock that have trapped dense concentrations of mammal bones which originally fell in or were washed in. Known as "fissure fills" of the Phosphorites de Quercy, these locations can yield thousands of extraordinarily preserved bones, but no complete skeletons, and there is no stratigraphic superposition, since each fissure could have opened at any time in the Eocene or Oligocene. Nevertheless, European paleontologists have made the best of this record and sorted out the complex history of fissure formation and filling in the Cenozoic. The overall picture was apparent to Hans Stehlin in 1909, when he described the differences between the Eocene and Oligocene faunas as *La Grande Coupure* ("the great cutoff")

Late Eocene faunas were still dominated by the endemic groups of the European archipelago, including the horse-like palaeotheres, dozens of endemic families of deer-like and pig-like artiodactyls, many endemic rodents, plus the last of the lemur-like primates, and both carnivorans and hyaenodont creodonts. By contrast, the Oligocene faunas were strikingly different. Some 60 percent of these Eocene endemic groups vanished. The new replacement groups were dominated by immigrants from Asia. The carnivorous mammals included leftover hyaenodonts, plus some of the earliest members of the bear, raccoon, weasel, and mongoose families, as well as beardogs and nimravids. Among perissodactyls, the last of the palaeotheres survived alongside true rhinoceroses, fleet hyracodont rhinos, and aquatic amynodont rhinos. The few surviving endemic artiodactyls had to compete with immigrant pig-like entelodonts, hippo-like anthracotheres, and a variety of advanced ruminants. The archaic Eocene rodents almost vanished, and were replaced by immigrant groups, including the aplodontids (represented today by the living sewellel), beavers, squirrels, dormice, and hamsters, plus the rabbits.

As we saw on other continents, the most striking differences are the loss of arboreal mammals (especially primates) and the replacement of obligatory leaf-eaters with mixed-feeders who are adapted to tougher vegetation and a mixture of leaves and scrub brush. Another way to detect this difference is the distribution of body sizes. If you plot all the herbivorous mammals in a fauna in order of increasing size, they form a distinctive plot known as a cenogram (Fig. 7.1). In a

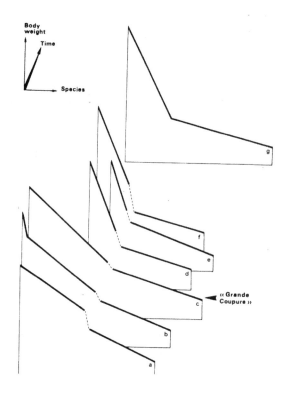

Figure 7.1. Cenogram of typical European faunas both before and after the Grande Coupure. Note that the warm tropical late Eocene faunas (top four diagrams) have more species (total length of cenogram), and a gradual slope without a major break, indicating the full range in body sizes from large animals (left) to small mammals (right). After the Grande Coupure, the cenogram (bottom two diagrams) is shorter (fewer species) and steeper, with a sharp drop in the middle representing the loss of mid-sized species (especially primates). (Modified from Legendre, 1992.)

tropical forest, the plot has a gentle slope with no break, because there is a high diversity of mammals of almost every body size. But in drier, scrubbier habitats, the shape of the cenogram is remarkably different. The slope is steeper (because large hoofed mammals are bigger), the line is shorter (because there are fewer total species), and there is a distinct break in the middle size range, reflecting the absence of medium-sized arboreal species such as the primates. Thus, the trends we have seen in qualitative form can be translated into graphical form as well.

Even though the faunal transformation was dramatic during the *Grande Coupure*, it is largely caused by wholesale immigration of new groups replacing the old ones, not by the dramatic climatic change itself. Floral analysis bears this out (Collinson and Hooker, 1987; Collinson, 1992). The floras of the late Eocene were dominated by evergreens, bald cypresses, and reed marshes, and those of the Oligocene were mixed deciduous/evergreen floras indicative of a warm-temperate climate. The pollen shows that the last of the tropical and subtropical vegetation disappeared, replaced by temperate plants and conifers (Collinson, 1992). Thus, there is clear evidence of cooling in Europe, but not of the drying trend seen in North America, probably because these small humid islands were not likely to undergo the drying seen in the center of the North American continent.

For a long time, European paleontologists correlated the *Grande Coupure* with the Eocene/Oligocene boundary (Cavelier, 1979; Savage and Russell, 1983) and a few European paleontologists insist on that correlation even today. But Hooker (1992) showed that the *Grande Coupure* occurred in the early Oligocene, consistent with what we now know about the early Oligocene global cooling event. When the Oligocene glaciers advanced, they caused a global drop in sea level as they locked up a huge amount of water. This sea level drop probably opened land bridges between the European archipelago and Asia, allowing new animal groups to immigrate in huge numbers.

Remodeling Mongolia

We have seen the climatic changes in the late Eocene and Oligocene that affected plants and animals in North America and Europe. So what kind of world did the indricotheres inhabit in the late Eocene and Oligocene? The Asian Oligocene fossil record is also excellent, not only in China and Mongolia, but also in many regions of the former Soviet Union, such as Kazakhstan and Georgia, as well as regions further south and east, such as India, Turkey, the Balkans, and Pakistan (Russell and Zhai, 1987; Lucas et al., 1998; Emry et al., 1998; Meng and McKenna, 1998). As we saw in Chapter 3, the early Oligocene in eastern Asia is known as the Shandgolian, after the famous locality of Hsanda Gol in Mongolia. In this unit, the faunas were very different from those of the late Eocene and also from the rest of the world. Meng and McKenna (1998) called this dramatic drop in diversity and transformation in the dominant faunas the "Mongolian Remodeling" (Fig. 7.2).

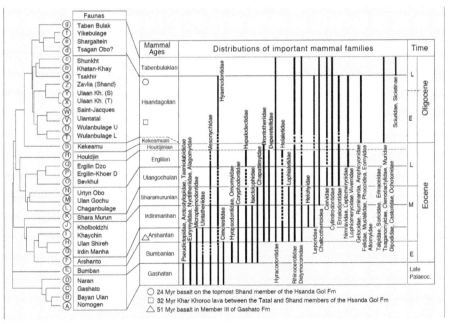

Figure 7.2. Dramatic drop in diversity of mammals in eastern Asia from the late Eocene (Ergilian) to the Oligocene (Hsanda Golian), which mirrors the transformation in Eocene-Oligocene mammal faunas from Europe and North America. (From Meng and McKenna, 1998; courtesy Meng Jin.)

It was truly a dramatic change. Many of the groups that dominated the late Eocene Ergilian landscape are gone, or nearly so. The giant brontotheres with battering-ram horns known as embolotheres were extinct (Fig. 3.6D), as were the last of the predatory giant mesonychids and most of the tapiroids that dominated in the late Eocene. Both hyracodont and true rhinos were common, as were deer and a variety of deer-like artiodactyls. Several new groups occur for the first time, including a variety of ruminant artiodactyls, several advanced rodents, plus a whole suite of advanced carnivorans: beardogs, weasels, and the first true cats. These shared the predator niche with the hyaenodonts, the cat-like nimravids, and the civets, which were already established.

Matthew (in Andrews, 1932, p. 107) commented:

> The character of the *Baluchitherium* fauna is peculiar as compared with most Tertiary mammal faunas, in the great abundance and variety of rodents and small carnivora, and scarcity of ungulates, especially artiodactyls. It represents probably a somewhat different facies from the badland faunas of western America, or the fissure and quarry faunas of western Europe. It may perhaps be a desert basin fauna.

Even though this list of characters seems similar to the early Oligocene faunas of other continents, there are important differences. The diversity of late Eocene large mammals is very reduced, with only rhinos and enteledonts occupying the large body size niche. This results in a cenogram (Fig. 7.3) that shows an even

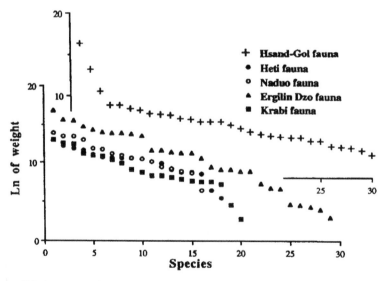

Figure 7.3. Cenograms for Asian Eocene-Oligocene faunas, showing the abrupt change in shape of the early Oligocene (Hsanda Golian) cenogram, with a steeper slope on the left (fewer species of very large mammals) and an abrupt change in slope where the mid-size (mainly arboreal) mammals would be found in the more forest-dominated Eocene faunas. (Modified from Ducrocq, 1993.)

more striking change between late Eocene and Oligocene faunas than does that of Europe (Fig. 7.1). There is a very steep slope on the left side of the Hsanda Gol curve (reflecting the handful of huge rhinos and entelodonts), then a very sharp break in slope as there were almost no mid-sized mammals (such as small ungulates and arboreal primates), and the remaining gradual slope of the plot on the right reflects the huge diversity of tiny rodents and lagomorphs. Such a steep slope, then abrupt break, is typical of more desert-like habitats today, where only a few large mammals (e.g., camels, wild asses) can survive, there are few medium-sized or tree-dwelling mammals, and the bulk of the mammalian diversity is in rodents or lagomorphs that can burrow.

The Hsanda Gol fauna is overwhelmingly dominated by rodents and lagomorphs, many with high-crowned teeth for eating gritty vegetation. According to Meng and McKenna (1998), the overall faunal composition suggests that the environment was very arid, with few tall trees for arboreal species. Almost all the land mammals are either adapted for feeding on the tops of the trees (like the indricotheres) or for living close to the ground and within the limited brush cover (most of the deer-like artiodactyls, entelodonts, and their predators, plus the great variety of ground-dwelling and burrowing rodents and rabbits). Leopold et al. (1992) studied the pollen from the Oligocene of China and concluded that much of the region was covered by a woody scrubland, with many arid-adapted plants, such as saltbush and *Ephedra,* or Mormon tea, as well an *Nitraria* (nitre bush, a salt-tolerant desert shrub). Trees would have been rare and concentrated in patches where groundwater could be found, such as in riparian habitats near river courses. In Siberia and Kazakhstan (Lucas et al., 1998), these trees included hardwoods such as oaks, as well as members of the walnut and birch families. In Mongolia and China, the Oligocene tree pollen is mostly from the birch and elm families, as well as oaks and other broad-leaved deciduous trees. These would have been the main fodder for the tree-feeding indricotheres. But in Pakistan, Martin et al. (2011) found evidence of a relatively dense but dry temperate to subtropical forest.

Li and Zheng (1995) found that the indricotheres lived in one of three climatic subregions of China, a semi-arid area dominated by scrub brush with dry lakes and dune deposits. The most common plant fossils are leaves of *Palibinia,* a fossil leaf taxon that appears to be highly adapted to the desert climate. The sedimentary rocks (Höck et al., 1999) suggest that the region also had abundant wind-blown dune sands, similar to those of the upper Oligocene Whitney and Arikaree rocks of North America, as well as gypsum and salt deposits formed in dry lakes.

Above all, Hsanda Gol and other Oligocene rocks of Asia are famous for *Paraceratherium.* As we saw in the Granger and Gregory (1935, 1936) reconstruction, this beast was 6 m (18 feet) tall at the shoulder, and probably weighed 15–20 tonnes. Its head was so high off the ground that it probably browsed the tops of trees, and it was taller than even modern elephants (Fig. 6.8). As we saw in Chapter 6, it probably lived a lot like elephants do today. There must have been small herds that roamed from place to place eating up all the vegetation (especially the high vegetation unreachable by small mammals). Each animal would have fed almost continuously to fuel its huge bulk, yet spent some of its day sleeping in the shade and keeping cool in water holes, roaming and eating mostly during the night. The herds would have been small, probably composed mostly mothers and their calves and their female kin. The adult indricotheres would have no problems with the small wolf-sized predators of the time, such as hyaenodonts. Like elephants, they would have herded to protect their young, since the young were the only ones who needed to fear the local predators. The young would have been born after a two-year gestation, and only one calf every few years.

In summary, although Shandgolian mammals show many families and genera in common with North America (especially the rhinos, entelodonts, anthracotheres, ruminants, plus many of the same weasels, beardogs, and nimravids), the fauna was much more arid-adapted than that of the Orellan in North America or the early Oligocene of Europe. In Asia, there were many more mammals bearing high-crowned teeth, and much floral evidence for grittier diets and scrubbier vegetation. Although many of these Asian groups migrated to Europe during the *Grande Coupure,* driving out the native fauna, Europe in the early Oligocene was still a much wetter and milder place than either Asia or North America.

Where Have All the Giants Gone?

Now that the ages and correlations of the Asian localities have been cleared up (Chapter 3), so that we no longer think indricotheres and amynodonts survived past the early Miocene in places like Dera Bugti, we can look at the indricotheres in the Oligocene of Asia and ask: what happened to them?

Naturally, such spectacular beasts invite no shortage of speculation about what caused their extinction, and nearly every paleontologist who has written about them has added to the stockpile of hypotheses. Henry Fairfield Osborn was famous for speculating that mammals showed "inadaptive evolution," evolving out of control until they were no longer able to survive. His most famous example was the supposed excessive size of the canines in saber-toothed cats. However, modern research showed that these canines occurred four separate times in mammals, so it must have worked fine for dozens of genera and species of different saber-tooths spanning the whole Cenozoic. Osborn also pointed to the supposedly "excessive" antlers of the "Irish elk," but Gould (1974) showed that these huge antlers were scaled appropriately for the largest deer ever known. Osborn (1929) also blamed the extinction of brontotheres on this idea of "inadaptive evolution" and "racial senescence," and Osborn (1923a) suggested that the same might have been true of indricotheres.

A modern version of this hypothesis was suggested by Janis and Carrano (1991), who suggested their extinction was due to their huge body size and presumed slow reproductive rate. However, indricotheres survived the entire Oligocene (34–23 Ma, or at least 11 million years) in some harsh desert-scrub environments and somehow managed to find a way to adapt and reproduce despite these less-than-ideal conditions. In addition, we saw in Chapter 6 that they were not that much larger than the biggest elephants and mammoths, also with slow reproductive rates, and these groups had many long-lived species that adapted to many different environments, and even to the rapidly changing environments of the Ice Ages.

How about changes in their environment? Gromova (1959) and Gabunia (1969) both suggested that the habitat for indricotheres changed as there were uplifts of the Himalayas or the Dinarides, causing rain shadows and desertification. However, we have already seen, the evidence from the Hsanda Gol and other Oligocene units suggests that the region was already an arid steppe-scrubland, and there is no evidence that it got any drier during the early Miocene (Yassamanov, 1985).

Putshkov and Kulczicki (1995) and Putshtkov (2001) suggested another possibility: competition from gomphothere mastodonts and their effects on the environment. As Norman Owen-Smith (1988) showed in his book *Megaherbivores,* modern African elephants have a drastic effect on their landscape. They are so strong and such voracious eaters that they rip large trees apart and tear up smaller trees and break huge gaps in dense vegetation. Studies have shown that where elephants are allowed to roam and feed freely, they radically change the environment. They can modify the dense forests and turn them into more open mixed habitats of brush, trees, and grasslands. The unmodified forests support one set of lower-diversity animals, but the elephant-modified landscapes support a much greater variety of animals that we consider the typical denizens of the African savanna, especially the variety of antelopes that live in varied habitats from open grassland to mixed scrub to dense scrub and forests. Owen-Smith (1987) called this the "keystone species" concept: the removal of one key species, like the landscape-changing elephants, can have cascading effects on the environment and other animals. Owen-Smith (1987) even suggested that it might explain the late Pleistocene extinction of the megamammals—once the mammoths vanished, they no longer modified the habitat, and many other mammals that had evolved in the habitats provided by mammoths went extinct as their own niches disappeared.

Putshkov and Kulczicki (1995) and Putshtkov (2001) point out that the early Miocene beds of Asia are marked by the first appearance of gomphothere mastodonts (Fig.7.4), which migrated out of Africa about 20 Ma when the closure and uplift Arabian Peninsula finally closed the opening to the eastern Mediterranean. They first appear in late Oligocene faunas of Asia about 20 Ma (Antoine et al., 2004) and in European mammal stage MN3b, or 19 Ma (Göhlich, 1999). The latest data indicate they appeared in North America by at least 16.2 Ma (Prothero et al., 2008). Indricotheres are last known from the latest Oligocene Tabenbulukian stage in Asia, and so far as I know, there is only one place where

Figure 7.4. Reconstruction of the head of the primitive mastodont Gomphotherium, *which left Africa in the latest Oligocene and spread across Eurasia and North America in the early Miocene. (Drawing by R. B. Horsfall; from Scott, 1930.)*

mastodonts and indricotheres co-occur: the Bugti beds (Antoine et al., 2004); they do not co-occur in Mongolia or China or Kazakhstan. And even in the Bugti beds, there is no co-occurrence between indricotheres and giant deinotheres (Fig. 6.10), with their downturned tusks, which would have been their most likely competition (Métais et al., 2009).

Putshkov and Kulczicki (1995) and Putshtkov (2001) suggest that gomphotheres and deinotheres, like living elephants, were capable of tremendous modification of the landscape, especially in breaking trees and their branches and stripping their bark. Lambert (1992) and Lambert and Shoshani (1998) showed that many primitive mastodonts, including the gomphotheres and the shovel-tuskers, show wear on their tusks indicative of stripping bark and breaking

Figure 7.5. Mounted skeleton of the gigantic beardog, Amphicyon ingens, *the largest predator on land during the middle Miocene in Eurasia and North America. (Photo courtesy L. Spoon.)*

branches and leaves. If they were anywhere near as destructive as modern elephants, they would have dramatically reduced the tree canopy on which indricotheres presumably fed.

Once indricothere populations (which probably had rather small population sizes, and required huge home ranges) were reduced due to habitat destruction by mastodont bulldozers, they might have been more vulnerable to other stresses, such as diseases, droughts, and especially new predators. The gigantic predatory hyaenodont *Hyainailouros* escaped from Africa in the early Miocene with the mastodonts, and the early Miocene also saw huge bear-sized predators like the beardog *Amphicyon ingens* (Fig. 7.5). Both of these predators were capable of taking large prey, like mastodont calves, and when they began to roam Asia as part of the Miocene invasion from Africa, they would have been formidable predators of indricothere calves as well (especially compared to the relatively small species of *Hyaenodon* and carnivorans that roamed Asia through the most of the Oligocene). And as we have already mentioned, there were giant 10 m (33 foot) crocodiles in the Bugti beds who were clearly major indricothere predators, since we have crocodile tooth marks on indricothere bones (P.-O. Antoine, pers. comm., 2011).

Of course, like any hypothesis about the prehistoric past, much of this is speculative, and we can never know for certain what caused a particular extinction. As Raup (1991) pointed out, we don't even know the actual cause of extinction of many animals, like the passenger pigeon, that roamed America just a century or two ago. So we must couch these hypotheses with caution and recognize that they are highly speculative. Still, the fact that indricotheres vanish just as both the mastodonts and their huge predators, as well as a phalanx of true rhinos and chalicotheres, first arrive in the early Miocene seems to be more than just a coincidence, especially since we can rule out climate and vegetational changes as well as older ideas like adaptive inflexibility.

Whatever the reason, the mighty indricotheres, the biggest mammals ever to roam the land, are gone forever. All we have is their incredibly huge skeletons to inspire awe and amazement and to give us a window into a world long gone.

Bibliography

Alexander, R. M. 1989. *Dynamics of Dinosaurs and Other Extinct Giants*. Columbia University Press, New York.

Andrews, R. C. 1932. *The New Conquest of Central Asia*. American Museum of Natural History, New York.

Antoine P.-O., S. M. I. Shah, I. U. Cheema, J.-Y. Crochet, D. de Franceschi, L. Marivaux, G. Métais, and J.-L. Welcomme. 2004. New remains of the baluchithere *Paraceratherium bugtiense* (Pilgrim, 1910) from the Late/latest Oligocene of the Bugti Hills, Balochistan, Pakistan. *Journal of Asian Earth Sciences* 24:71–77.

Antoine, P.-O., L. Karadenizli, G. Sarac, and S. Sen. 2008. A giant rhinocerotoid (Mammalia, Perissodactyla) from the Late Oligocene of north-central Anatolia (Turkey). *Zoological Journal of the Linnaean Society* 152:581–592.

Beliajeva, E. 1959. Sur la decouverte de rhinoceros Tertiares anciens dans la province maritime de l'U.R.S.S. *Vertebrata PalAsiatica* 3:81–91.

Berkey, C., and F. Morris. 1927. *Geology of Mongolia. Natural History of Central Asia, Volume II*. American Museum of Natural History, New York, New York.

Bodylevskaya, I. V. 2008. *Academician A. A. Borissiak and the Paleontological Institute during the Second World War, 1941–1943*. Paleontological Institute, Moscow.

Borisov, B. A. 1963. Stratigrafiya verkhnevo Mela i Paleyoegyen-Neogyena Zaisanskoi vpadiny. *Trudy VSEGEI i Gosudarstveny geologischekiy Kominyeta, Novye Seiy* 94:11–75.

Borissiak, A. A. 1915. Ob indrikoterii (*Indricotherium* n.g.) *Geologik Vestnik* 1(3):131–134.

Borissiak, A. A. 1923. O rod *Indricotherium* n.g. (sem. Rhinocerotidae). *Zapinski Russiches Akademik Nauk* (8) 35(6):1–128.

Bown, T. M., and M. J. Kraus. 1981. Lower Eocene alluvial paleosols (Willwood Formation, northwest Wyoming, USA) and their significance for paleoecology, paleoclimatology, and basin analysis. *Palaeogeography, Palaeoclimatology, Palaeoecology* 34:1–30.

Bown, T. M., and M. J. Kraus. 1987. Integration of channel and floodplain suites: 1, Developmental sequence and lateral relations of alluvial paleosols. *Journal of Sedimentary Petrology* 57:587–601.

Bryant, D., and M. C. McKenna. 1995. Cranial anatomy and phylogenetic position of *Tsaganomys altaicus* (Mammalia, Rodentia) from the Hsanda Gol Formation (Oligocene), Mongolia. *American Museum Novitates* 3156:1–42.

Budantsev, L. Y. 1992. Early stage of formation and dispersal of the temperate flora in the Boreal region. *The Botanical Review* 58:1–48.

Calandra, I., U. B. Göhlich, and G. Merceron. 2008. How could sympatric megaherbivores coexist? Example of niche partitioning within a proboscidean community from the Miocene of Europe. Naturwissenschaften 95:831–838.

Cavelier, C. 1979. La limite Eocène-Oligocène en Europe occidentale. *Sci. Géol. Inst. Géol. Strasbourg, (Mém.)* 54:1–280.

Chow, M., and C. Chiu. 1963. A new genus of giant rhinoceros from Oligocene of Inner Mongolia. *Vertebrata PalAsiatica* 7:230–239.

Chow, M., and C. Chiu. 1964. An Eocene giant rhinoceros. *Vertebrata PalAsiatica* 8:264–267.

Christiansen, P. 2004. Body size in proboscideans, with notes on elephant metabolism. Zoological Journal of the Linnean Society 140:523–549.

Clauss, M., R. Frey, B. Kiefer, M. Lechner-Doll, W. Loehlein, C. Polster, G. E. Rössner, and W. J. Streich. 2003. The maximum attainable body size of herbivorous mammals: morphophysiological constraints on foregut, and adaptations of hindgut fermenters. *Oecologia* 136:14–27.

Codrea, V. 2000. *Rinoceri si Tapiri Tertiari din Romania*. Presa Universitara Clujeana, Cluj.

Collinson, M. E. 1983. Palaeofloristic assemblages and palaeoecology of the lower Oligocene Bembridge Marls, Hamstead Ledge, Isle of Wight. Botanical Journal of the Linnean Society 86:177–225.

Collinson, M. E. 1992. Vegetational and floristic changes around the Eocene/Oligocene boundary in western and central Europe, pp. 437–450, *in* Prothero, D. R., and W. A. Berggren (eds.), *Eocene-Oligocene Climatic and Biotic Evolution*. Princeton University Press, Princeton, New Jersey.

Collinson, M. E., and J. J. Hooker. 1987. Vegetational and mammalian faunal changes in the early Tertiary of southern England, pp. 295–304, *in* Friis, E. M., W. G. Chaloner, and P. R. Crane (eds.), *The Origins of Angiosperms and their Biological Consequences*. Cambridge University Press, Cambridge.

Creber, G. T., and W. G. Chaloner. 1984. Climatic indications from growth rings in fossil woods, pp. 49–74, *in* P. J. Brenchley, (ed.), *Fossils and Climate*. John Wiley and Sons, New York.

Damuth, J. 1990. Problems in estimating body masses of archaic ungulates using dental measurements, pp. 229–253, *in* Damuth, J., and B. J. MacFadden, (eds.), *Body Size in Mammalian Paleobiology*. Cambridge University Press, Cambridge.

Dashzeveg, D. 1993. Asynchronism of the main mammalian faunal events near the Eocene-Oligocene boundary. *Tertiary Research* 14:141–143.

Daxner-Höck, G., and D. Badamgarav. 2007. Geological and stratigraphic setting. *Annalen des Naturhistorischen Museums in Wien* 108A:1–24.

Dayan, T., D. Wool, and D. Simberlofff. 2002. Varation and covariation of skulls and teeth: modern carnivores and the interpretation of fossil mammals. *Paleobiology* 28:508–526.

du Toit, J. T. 1990. Home range-body mass relations: a field study on African browsing ruminants. *Oecologia* 85:301–303.

Ducrocq, S. 1993. Mammals and stratigraphy in Asia: is the Eocene-Oligocene boundary at the right place? *Compte Rendus Academie de Science Paris, Serie* II, 316:419–426

Durden, C. J. 1966. Oligocene lake deposits in central Colorado and a new fossil insect locality. *Journal of Paleontology* 40:215–219.

Economos, A. C. 1981. The largest land mammal. *Journal of Theoretical Biology* 89:211–215.

Emry, R. J., S. G. Lucas, L. Tyutkova, and B. Wang. 1998. The Ergilian-Shandgolian (Eocene-Oligocene) transition in the Zaysan Basin, Kazakhstan. *Bulletin of Carnegie Museum of Natural History* 34:298–312.

Evanoff, E., D. R. Prothero, and R. H. Lander. 1992. Eocene-Oligocene climatic change in North America: the White River Formation near Douglas, east-central Wyoming, pp. 116–130, *in* Prothero, D. R. and W. A. Berggren (eds.), *Eocene-Oligocene Climatic and Biotic Evolution.* Princeton University Press, Princeton, New Jersey.

Evanoff, E., W. C. McIntosh, and P. C. Murphey. 2001. Stratigraphic summary and [40]Ar/[39]Ar geochronology of the Florissant Formation, Colorado, pp. 1–16, *in* Evanoff, E., K. M. Gregory-Wodzicki, and K. R. Johnson (eds.), Fossil flora and stratigraphy of the Florissant Formation, Colorado: *Proceedings of the Denver Museum of Nature and Science, Series* 4, no. 1.

Farlow, J. O., I. D. Coroian, and J. R. Foster. 2010. Giants on the landscape: modeling the abundance of megaherbivorous dinosaurs of the Morrison Formation (Late Jurassic, western USA). *Historical Biology* 22:403–429.

Forster Cooper, C. 1911. *Paraceratherium bugtiense,* a new genus of Rhinocerotidae from Bugti Hills of Baluchistan—preliminary notice. *Annual Magazine of Natural History* 8(8):711.

Forster Cooper, C. 1913a. *Thaumastotherium osborni,* a new genus of perissodactyls from the Upper Oligocene deposits of the Bugti Hills of Baluchistan—preliminary notice. *Annual Magazine of Natural History* 8(12): 367–381.

Forster Cooper, C. 1913b. Correction of generic name [*Thaumastotherium* to *Baluchitherium*]. *Annual Magazine of Natural History* 8(12):504.

Forster Cooper, C. 1923a. *Baluchitherium osborni* (?syn. *Indricotherium turgaicum* Borissiak). *Philosophical Transactions of the Royal Society of London* B 50:35–66.

Forster Cooper, C. 1923b. On the skull and dentition of *Paraceratherium bugtiense:* a genus of aberrant rhinoceroses from the lower Miocene deposits of

Dera Bugti. *Philosophical Transactions of the Royal Society of London* B 212:369–394.

Forster Cooper, C. 1934. The extinct rhinoceroses of Baluchistan. *Philosophical Transactions of the Royal Society of London* B 223:569–616.

Fortelius, M., and J. Kappelman. 1993. The largest land mammal ever imagined. *Zoological Journal of the Linnean Society* 108(1):85–101.

Fortelius, M., and N. Solounias. 2000. Functional characterization of ungulate molars using the abrasion-attrition wear gradient: a new method for reconstructing paleodiets. *American Museum Novitates* 3301:1–36.

Gabunia, L. 1955. O svoebraznom pred stavitele Indricotheriidae iz Oligotsena Gruzii. *DokladyAkademie Naul Armeniya SSSR*. 21:177–181.

Gabunia, L. 1964. *Benarskaya Fauna Oligotsenovykh pozvonochnykh*. Metsniereba Press, Tblisi.

Gabunia, L. 1966. Sur les Mammifères oligocènes du Caucase. *Bulletin de la Société geologique de France* 7:857–869.

Gabunia, L. 1969. *Extinction of Ancient Reptiles and Mammals*. Metsniereba, Tblisi.

Gingerich, P. D. 1990. Predictions of body mass in mammalian species from long bone lengths and diameters. *Contributions from the Museum of Paleontology, University of Michigan* 28:79–92.

Göhlich, U. B. 1999. Order Proboscidea, pp. 157–168, *in* Rössner, G. E., and K. Heissig(eds.), *The Miocene Land Mammals of Europe*. Verlag Dr. Friedrich Pfeil, Munich.

Gould, S. J. 1974. Origin and Function of 'Bizarre' Structures—Antler Size and Skull Size in 'Irish Elk'. *Megaloceros giganteus. Evolution* 28(2):191–220.

Gradstein, F., J. G. Ogg, and A. Smith. 2004. *A Geologic Time Scale 2004*. Cambridge University Press, Cambridge.

Graham, A. 1999. *Late Cretaceous and Cenozoic History of North American Vegetation*. Oxford University Press, Oxford.

Granger, W., and W. K. Gregory. 1935. A revised restoration of the skeleton of *Baluchitherium*, gigantic fossil rhinoceros of central Asia. *American Museum Novitates* 787:1–4.

Granger, W., and W. K. Gregory. 1936. Further notes on the gigantic extinct rhinoceros *Baluchitherium* from the Oligocene of Mongolia. *Bulletin of the American Museum of Natural History* 72:1–73.

Greenwood, D. R., P. T. Moss, A. I. Rowett, A. J. Vadala, and R. J. Keefe. 2003. Plant communities and climate change in southeastern Australia during the early Paleogene. *Geological Society of America Special Paper* 369:365–380.

Gregory, W. K., and H. J. Cook. 1928. New materials for the study of evolution: A series of primitive fossil rhinoceros skulls (*Trigonias*) from the Lower Oligocene of Colorado. *Proceedings of the Colorado Museum of Natural History* 7(1):1–32.

Gromova, V. 1959. Gigantskie norosogi [Gigantic rhinoceroses]. *Trudy Paleontologik Institute Akademik Nauk SSR* 71:1–164.

Guo, S. X. 1985. Preliminary interpretation of Tertiary climate by using megafossil floras in China. *Palaeotologia Cathayana* 2:169–175.

Heissig, K. 1979. Die hypothetische Rolle Südosteuropas bei den Säugetierwanderungen im Eozän und Oligozän. *Neues Jahrbuch für Geologie und Paläontologie* 2:83–96.

Höck, V., G. Daxner-Höck, H. P. Schmid, D. Badamgarav, W. Frank, G. Furtmüller, O. Montag, R. Barsbold, Y. Khand, and J. Sodov. 1999. Oligocene-Miocene sediments, fossils and basalts from the Valley of Lakes (Central Mongolia), An integrated study. *Mitteilungen der Österreichischen Geologischen Gesellschaft* 90:83–125.

Holbrook, L. T., and S. G. Lucas. 1997. A new genus of rhinocerotoid from the Eocene of Utah and the status of North American "*Forstercooperia.*" *Journal of Vertebrate Paleontology* 17:384–396.

Hooker, J. J. 1992. British mammalian paleocommunities across the Eocene-Oligocene transition and their environmental implications, pp. 494–515, *in* Prothero, D. R., and W. A. Berggren (eds.), *Eocene-Oligocene Climatic and Biotic Evolution*. Princeton University Press, Princeton, New Jersey.

Hutchison, J. H. 1982. Turtle, crocodilian and champsosaur diversity changes in the Cenozoic of the north-central region of the western United States. *Palaeogeography, Palaeoclimatology, Palaeoecology* 37:149–164.

Hutchison, J. H. 1992. Western North American reptile and amphibian record across the Eocene/Oligocene boundary and its climatic implications, pp. 451–463, *in* Prothero, D. R., and W. A. Berggren (eds.), *Eocene-Oligocene Climatic and Biotic Evolution*. Princeton University Press, Princeton, New Jersey.

Janis, C. 1976. The evolutionary strategy of the Equidae and the origins of rumen and caecal digestion. *Evolution* 30:757–774.

Janis, C. M., and M. Carrano. 1991. Scaling of reproductive turnover in archosaurs and mammals: why are large terrestrial mammals so rare? *Annales Zoologica Fennica* 28:201–206.

Kirkaldy, G. W. 1908. Memoir on a few heteropterous Hemiptera from eastern Australia. *Proceedings of the Linnean Society of New South Wales* 32:768–788.

Kraatz, B. P., and J. H. Geisler. 2010. Eocene-Oligocene transition in Central Asia and its effects on mammalian evolution. *Geology* 38:111–114.

Krishtalka, L. 1989. The naming of the shrew, pp. 28–37, in *Dinosaur Plots*. William Morrow, New York.

Kurtén, B. 1953. On variation and population dynamics of fossil and recent mammal populations. *Acta Zoologica Fennica* 76:1–122.

Lambert, W. D. 1992. The feeding habits of the shovel-tusked gomphotheres: evidence from tusk wear patterns. *Paleobiology* 18:133–147.

Lambert, W. D., and J. Shoshani. 1998. Proboscidea, pp. 606–622, *in* C. Janis, K. M. Scott, and L. Jacobs (eds.), *Evolution of Tertiary Mammals of North America*. Cambridge University Press, Cambridge.

Lande, R. 1977. On comparing coefficients of variation. *Systematic Zoology* 26:214–217.

Leopold, E. B., L. Gengwu, and S. Clay-Poole. 1992. Low-biomass vegetation in the Oligocene? pp. 399–420, *in* Prothero, D. R., and W. A. Berggren (eds.), *Eocene-Oligocene Climatic and Biotic Evolution*. Princeton University Press, Princeton, New Jersey.

Lewis, G. E. 1944. Guy E. Pilgrim, 1875–1943 (obituary biography). *American Journal of Science* 242:105–106.

Li, H.-M., and Y.-H. Zheng. 1995. Palaeogene flora, pp. 455–505, *in* X.-X. Li (ed.), *Fossil Floras of China through the Geological Ages*. Guangdong Science and Technology Press, Guangdong.

Lucas, S. G., and B. U. Bayshashov. 1996. The giant rhinoceros *Paraceratherium* from the late Oligocene at Aktau Mountain, southeastern Kazakhstan, and its biochronologic significance. *Neues Jahrbuch für Geologie und Paläontologie Monatshefte* 1996:539–548.

Lucas, S. G., and J. C. Sobus. 1989. The systematics of indricotheres, pp. 358–378, in Prothero, D. R., and R. M. Schoch (eds.), *The Evolution of Perissodactyls*. Oxford University Press, New York.

Lucas, S. G., and O. G. Bendukidze.1997. Proboscidea (Mammalia) from the early Miocene of Kazakhstan. *Neues Jahrbuch für Geologie und Paläontologie Monatshefte* 1997:659–673.

Lucas, S. G., and R. J. Emry. 1996a. Early Record of Indricothere (Mammalia: Perissodactyla: Hyracondontidae) from the Aral Sea Region of Western Kazakhstan. *Proceedings of the Biological Society of Washington* 109(2):391–396.

Lucas, S. G., and R. J. Emry. 1996b. Late Eocene Entelodonts (Mammalia: Artiodactyla) from Inner Mongolia, China. *Proceedings of the Biological Society of Washington* 109(2):397–405.

Lucas, S. G., and R. J. Emry. 1996c. Biochronological Significance of Amynodontidae (Mammalia, Perissodactyla) from the Paleogene of Kazakhstan. *Journal of Paleontology* 760(4):691–696

Lucas, S. G., and R. J. Emry. 1999. Taxonomy and biochronological significance of *Paraentelodon*, a giant entelodont (Mammalia, Artiodactyla) from the late Oligocene of Eurasia. *Journal of Vertebrate Paleontology* 19:160–168.

Lucas, S. G., E. G. Kordikova, and R. J. Emry. 1998. Oligocene stratigraphy, sequence stratigraphy, and mammalian biochronology north of the Aral Sea, western Kazakhstan. *Bulletin of Carnegie Museum of Natural History* 34:313–348.

Lucas, S. G., R. M. Schoch, and E. Manning. 1981. The systematics of *Forstercooperia*, a middle to late Eocene hyracodontid (Perissodactyla: Rhinocerotoidea) from Asia and western North America. *Journal of Paleontology* 55:826–841.

Lucas, S. G., B. U. Bayshashov, L. A. Tyutkova, A. K. Zhamangara, and B. Z. Aubekerov. 1997. Mammalian biochronology of the Paleogene-Neogene boundary at Aktau Mountain, eastern Kazakhstan. *Paläontologische Zeitschrift* 71:305–314.

Macdonald, J. R. 1963. The Miocene faunas from the Wounded Knee area of western South Dakota. *Bulletin of the American Museum of Natural History* 125:139–238.

Marivaux, L., P.-O. Antoine, S. F. H. Baqri, M. Benammi, Y. Chaimanee, J.-Y. Crochet, D. Franchesci, N. Iqbal, J.-J. Jaeger, G. Metais, G. Roohi, and J.-L. Welcomme. 2005. Anthropoid primates from the Oligocene of Pakistan (Bugti Hills): Data on early anthropoid evolution and biogeography. *Proceedings of the National Academy of Sciences (USA)* 102:8436–8441.

Martin, C., I. Bentaleb, P.-O. Antoine. 2011. Pakistan mammal tooth stable isotopes show paleoclimatic and paleoenvironmental changes since the early Oligocene. Palaeogeography, Palaeoecology, Palaeoclimatology 311(1):19–29.

Matthew, W. D. 1931. Critical observations on the phylogeny of the rhinoceroses. *University of California Publications, Bulletin of the Department of Geological Sciences* 20:1–9.

Matthew, W. D., and W. Granger. 1923. The fauna of the Houldjin gravels. *American Museum Novitates* 97:1–6.

Mayr, E. 1969. *Principles of Systematic Zoology.* McGraw-Hill, New York.

McKenna, M. C. 1980. Eocene paleolatitude, climate and mammals of Ellesmere Island. *Palaeogeography, Palaeoclimatology, Palaeoecology* 30:349–362.

McKenna, M. C. 1983. Holarctic landmass rearrangement, cosmic events, and Cenozoic terrestrial organisms. *Annals of the Missouri Botanical Garden* 70:459–489.

McKenna, M. C., and S. K. Bell. 1997. *Classification of Mammals Above the Species Level.* Columbia University Press, New York.

Mehrotra, R. C. 2003. Status of plant megafossils during the early Paleogene in India. *Geological Society of America Special Paper* 369:413–423.

Meiri, S., T. Dayan, and D. Simberloff. 2005. Variability and correlations in carnivore crania and dentition. *Functional Ecology* 19:337–343.

Mellett, J. S. 1968. The Oligocene Hsanda Gol Formation, Mongolia: a revised faunal list. *American Museum Novitates* 2318:1–16.

Mellett, J. S. 1982. Body size, diet, and scaling factors in large carnivores and herbivores. *Proceedings of the Third North American Paleontological Convention* 1:371–376.

Meng, J., and M. C. McKenna. 1998. Faunal turnovers of Paleogene mammals from the Mongolian plateau. *Nature* 394:364–367.

Métais, G., P.-O. Antoine, S. R. H. Baqri, L. Marivaux, J.-L. Welcomme. 2009. Lithofacies, depositional environments, regional biostratigraphy, and age of the Chitarwata Formation in the Bugti Hills, Balochistan, Pakistan. *Journal of Asian Earth Sciences* 34:154–167.

Meyer, H. W. 2003. *The Fossils of Florissant:* Smithsonian Institution Press, Washington, D.C.

Meyer, H. W., and S. R. Manchester. 1997. Revision of the Oligocene Bridge Creek floras of Oregon. *University of California Publication in Geological Sciences* 141:1–195.

Miao D.-S., C.-K. Li, and Y.-Q. Wang. 2010. Minchen Chow's academic marathon from the Bighorn Basin to the Nanxiong Basin. *Vertebrata PalAsiatica* 48:281–284.

Mihlbachler, M. C. 2008. Species taxonomy, phylogeny, and biogeography of the Brontotheriidae (Mammalia: Perissodactyla). *Bulletin of the American Museum of Natural History* 311:1–473.

Mihlbachler, M. C., S. G. Lucas, and R. J. Emry. 2004. The holotype specimen of *Menodus giganteus,* and the "insoluble" problem of Chadronian brontothere taxonomy. *Bulletin of the New Mexico Museum of Natural History* 26:129–136.

Myers, J. 2003. Terrestrial Eocene-Oligocene vegetation and climate in the Pacific Northwest, pp. 171–188, *in* Prothero, D. R., L. C. Ivany, and E. A. Nesbitt (eds.), *From Greenhouse to Icehouse: The Marine Eocene-Oligocene Transition.* Columbia University Press, New York.

Nikolov, I., and K. Heissig. 1985. Fossile Säugetiere as dem Obereozän und Unteroligozän Bulgariens und ihre Bedeutung für die Paläogeographie. *Mittelungen des Bayerische Staatsammlung für Paläontologie und Historische Geologie* 25:61–79.

Olson, E. C. 1990. *The Other Side of the Medal: A Paleontologist Reflects on the Art and Serendipity of Science.* McDonald and Woodward Publishing, New York.

Orliac, M. J., P.-O. Antoine, G. Roohi, and J.-L. Welcomme. 2010. Suoidea (Mammalia, Cetartiodactyla) from the early Oligocene of the Bugti Hills, Balochistan, Pakistan. *Journal of Vertebrate Paleontology* 30:1300–1305.

Osborn, H. F. 1923a. *Baluchitherium grangeri,* a giant hornless rhinoceros from Mongolia. *American Museum Novitates* 78:1–15.

Osborn, H. F. 1923b. The extinct giant rhinoceros *Baluchitherium* of western and central Asia. *Natural History* 23(May–June):209–228.

Osborn, H. F. 1929. The titanotheres of ancient Wyoming, Nebraska, and South Dakota. *U.S. Geological Survey Monograph* 55:1–953 (2 vols.).

Owen-Smith, N. R. 1987. Pleistocene extinctions: the pivotal role of megaherbivores. *Paleobiology* 13:351–362.

Owen-Smith, N. R. 1988. *Megaherbivores: The Influence of Very Large Body Size on Ecology.* Cambridge University Press, Cambridge.

Pavlova, M. 1922. *Indricotherium transouralicum,* n.s. provenant du district de Tourgay. *Bulletin of the Society Naturalik Moscou, Section of Geology* 2(31):95–116.

Pilgrim, G. 1908. The Tertiary and post-Tertiary fresh-water deposits of Baluchistan and Sind, with notices of new vertebrates. *Records of the Geological Survey of India* 37:139–166.

Pilgrim, G. 1910. Notices of new mammalian genera and species from the territories of India. *Records of the Geological Survey of India* 40:63–71.

Plavcan, J. M., and D. A. Cope. 2001. Metric variation and species recognition in the fossil record. *Evolutionary Anthropology* 10:204–222.

Polly, P. D. 1998. Variability in mammalian dentitions: size-related bias in the coefficient of variation. *Biological Journal of the Linnaean Society* 64:83–99.

Preston, D. J. 1986. *Dinosaurs in the Attic: An Excursion into the American Museum of Natural History.* St. Martin's Press, New York.

Prothero, D. R. 1994. *The Eocene-Oligocene Transition: Paradise Lost.* Columbia University Press, New York.

Prothero, D. R. 1996a. Hyracodontidae, pp. 634–645, *in* Prothero, D. R., and R. J. Emry (eds.), *The Terrestrial Eocene-Oligocene Transition in North America.* Cambridge University Press, Cambridge.

Prothero, D. R. 1996b. Camelidae, pp. 591–633, *in* Prothero, D. R., and R. J. Emry (eds.), *The Terrestrial Eocene-Oligocene Transition in North America.* Cambridge University Press, Cambridge.

Prothero, D. R. 1999. Does climatic change drive mammalian evolution? *GSA Today* 9(9):1–5.

Prothero, D. R. 2005. *The Evolution of North American Rhinoceroses.* Cambridge University Press, Cambridge.

Prothero, D. R. 2006. *After the Dinosaurs: The Age of Mammals.* Indiana University Press, Bloomington.

Prothero, D. R. 2008. Magnetic stratigraphy of the Eocene-Oligocene floral transition in western North America. *Geological Society of America Special Paper* 435:71–87.

Prothero, D. R. 2009. *Catastrophes!* Johns Hopkins University Press, Baltimore, Maryland.

Prothero, D. R., and C. C. Swisher III, 1992. Magnetostratigraphy and geochronology of the terrestrial Eocene-Oligocene transition in North America, pp. 46–74, *in* Prothero, D. R., and W. A. Berggren (eds.), *Eocene-Oligocene Climatic and Biotic Evolution.* Princeton University Press, Princeton, New Jersey.

Prothero, D. R., and F. Sanchez. 2004. Magnetic stratigraphy of the upper Eocene Florissant Formation, Teller County, Colorado: New Mexico. *Museum of Natural History and Science Bulletin* 26:143–147.

Prothero, D. R., and F. Schwab. 2012. *Sedimentary Geology* (3rd ed.). W. H. Freeman, New York.

Prothero, D. R., and R. J. Emry. 2004. The Chadronian, Orellan, and Whitney and North American land mammal ages, pp. 156–168, in M. O. Woodburne (ed.), *Late Cretaceous and Cenozoic Mammals of North America.* Columbia University Press, New York.

Prothero, D. R., and R. M. Schoch. 1989. The origin and evolution of perissodactyls: a summary and synthesis, pp. 504–529, in D. R. Prothero, and R. M.

Schoch (eds.), The Evolution of Perissodactyls. Oxford University Press, New York.

Prothero, D. R., and R. M. Schoch. 2002. *Horns, Tusks, and Flippers: The Evolution of Hoofed Mammals*. Johns Hopkins University Press, Baltimore, Maryland.

Prothero, D. R., and T. H. Heaton, 1996. Faunal stability during the early Oligocene climatic crash. *Palaeogeography, Palaeoclimatology, Palaeoecology* 127:239–256.

Prothero, D. R., and W. A. Berggren (eds.), 1992. *Eocene-Oligocene Climatic and Biotic Evolution*. Princeton University Press, Princeton, New Jersey.

Prothero, D. R., C. Guérin, and E. Manning, 1989. The history of the Rhinocerotoidea, pp. 322–340, *in* Prothero, D. R., and R. M. Schoch (eds.), *The Evolution of Perissodactyls*. Oxford University Press, New York.

Prothero, D. R., E. Manning, and C. B. Hanson, 1986. The phylogeny of the Rhinocerotoidea (Mammalia, Perissodactyla). *Zoological Journal of the Linnean Society of London* 87:341–366.

Prothero, D. R., E. B. Davis, and S. S. B. Hopkins. 2008. Magnetic stratigraphy of the early Miocene (late Hemingfordian) Massacre Lake fauna, northwest Nevada, and its implications for the "Proboscidean Datum" in North America. *New Mexico Museum of Natural History and Science Bulletin* 44:239–246.

Putshkov, P. V. 2001. "Proboscidean agent" of some Tertiary megafaunal extinctions. *The World of Elephants*. International Congress, Rome, pp. 133–136.

Putshkov, P. V., and A. H. Kulczicki. 1995. The early Miocene drama: mastodonts against indricotheres. *Vestnik Zoologii* 1995(6):54–62.

Qiu, Z.-X., and B.-Y. Wang. 2007. Paracerathere fossils of China. *Paleontologica Sinica* 193(29):1–396.

Radinsky, L. B. 1966. The families of the Rhinocerotoidea (Mammalia, Perissodactyla). *Journal of Mammalogy* 47:631–639.

Radinsky, L. B. 1967. A review of the rhinocerotoid family Hyracodontidae (Perissodactyla). *Bulletin of the American Museum of Natural History* 136:1–46.

Radinsky, L. B. 1969. The early evolution of the Perissodactyla. *Evolution* 23:308–328.

Rainger, R. 1991. *An Agenda for Antiquity: Henry Fairfield Osborn and Vertebrate Paleontology at the American Museum of Natural History, 1890–1935*. University of Alabama Press, Tuscaloosa, Alabama.

Raup, D. M. 1991. *Extinctions: Bad Genes or Bad Luck?* W. W. Norton, New York.

Raza, S. M., and G. E. Meyer. 1984. Early Miocene geology and paleontology of the Bugti Hills. *Geological Survey of Pakistan* 11:43–63.

Retallack, G. J. 1983. Late Eocene and Oligocene paleosols from Badlands National Park, South Dakota. *Geological Society of America Special Paper* 193.

Retallack, G. J., W. N. Orr, D. R. Prothero, R. A. Duncan, P. R. Kester, and C. P.

Ambers. 2004. Eocene-Oligocene extinctions and paleoclimatic change near Eugene, Oregon. Geological Society of America Bulletin 116:817–839.

Romero, E. J. 1986. Paleogene phytogeography and paleoclimatology of South America. *Annals of the Missouri Botanical Garden* 73:449–461.

Russell, D. E., and R. J. Zhai. 1987. The Paleogene of Asia: Mammals and Stratigraphy. *Mémoires du Muséum National d'Histoire Naturelle, Sciences de la Terre* 52:1–488.

Sahni, A., and S. K. Khare. 1972. Additional Eocene mammals from the Subathu Formation of Jammu and Kashmir. *Journal of the Paleontological Society of India* 17:31–49.

Sander, P. M., A. Christian, M. Clauss, M. Fechner, C. T. Gee, E.-M. Griebeler, H.-C. Gunga, J. Hummel, H. Mallison, St. F. Perry, H. Preuschoft, O. W. M. Rauhut, C. Remes, T. Thutken, O. Wings, and U. Witzel. 2011. Biology of the sauropod dinosaurs: the evolution of gigantism. *Biological Reviews* 86:117–155.

Savage, D. E., and D. E. Russell. 1983. *Mammalian Paleofaunas of the World*. Addison-Wesley, Reading, Massachusetts.

Savage, R. J. G., and M. R. Long. 1986. *Mammal Evolution: An Illustrated Guide*. Facts on File Publications, New York.

Scheele, W. 1955. *The First Mammals*. World Press, New York.

Sen, S., P.-O. Antoine, B. Varol, T. Ayyildiz, and K. Sözeri. 2011. Giant rhinoceros *Paraceratherium* and other vertebrates from Oligocene and middle Miocene deposits of the Kagizman-Tuzluca basin, eastern Turkey. *Naturwissenschaften* 98:407–423.

Simpson, G. G. 1961. *Principles of Animal Taxonomy*. Columbia University Press, New York.

Simpson, G. G., A. Roe, and R. C. Lewontin. 1960. *Quantitative Zoology*. Harcourt Brace and World, New York.

Sokal, R. R., and F. J. Rohlf. 1994. *Biometry* (3rd ed.). W. H. Freeman, New York.

Solounias, N., and G. M. Semprebon. 2002. Advances in the reconstruction of ungulate ecomorphology and application to early fossil equids. *American Museum of Natural History Novitates* 3366: 1–49.

Spassov, N. 1989. On the taxonomic status of *Indricotherium* Borissiak and giant rhinocerotoids-indricotheres (Perissodactyla). *Comptes Rendus de l'Academie bulgare des Sciences* 42:61–64.

Swisher, C. C., III, and D. R. Prothero. 1990. Single-crystal ^{40}Ar/^{39}Ar dating of the Eocene-Oligocene transition in North America. *Science* 249:760–762.

Teilhard de Chardin, P. 1926. Mammifères Tertiares de Chine et de Mongolie. *Annals of Paleontologie* 15:1–51.

Tiwari, B. N. 2003. A late Eocene *Juxia* (Perissodactyla, Hyracodontidae) from Liyan Molasse, eastern Ladakh, India. *Journal of the Paleontological Society of India* 48:103–113.

Tong, Y., S. Zheng, and Z. Qiu. 1995. Cenozoic mammal ages of China. *Vertebrata PalAsiatica* 33:290–314.

Wang, B.-Y. 1992. The Chinese Oligocene: a preliminary review of mammalian localities and local faunas, pp. 529–547, in Prothero, D. R., and W. A. Berggren (eds.), *Eocene-Oligocene climatic and biotic evolution*. Princeton University Press, Princeton, New Jersey.

Wang, B.-Y. 2007. Late Eocene lagomorphs from Nei Mongol, China. *Vertebrata PalAsiatica* 45:43–58.

Wang, B.-Y., Z.-X. Qiu, Q.-Z. Zhang, L.-J. Wu, and P.-J. Ning. 2009. Large mammals found from Houldjin Formation near Erenhot, Nei Mongol, China. *Vertebrata PalAsiatica* 47:85–110.

Wang, J. 1976. A new genus of Forstercooperiinae from the late Eocene of Tongbo, Henan. *Vertebrata PalAsiatica* 14:104–111.

Wang, Y., and T. Deng. 2005. A 25 m.y. isotopic record of paleodiet and environmental change from fossil mammals and paleosols from the NE margin of the Tibetan Plateau. *Earth and Planetary Science Letters* 236:322–338.

Watson, D. M. S. 1950. Clive Forster-Cooper, 1880–1947 (obituary biography). *Biographical Memoirs of the Fellows of the Royal Society of London* 7:82–93.

Welcomme, J-L., and L. Ginsburg. 1997. The evidence of an Oligocene presence in the Bugti area (Balouchistan, Pakistan). *Comptes es rendus de l'Academie des Sciences* 325:999–1004.

Welcomme, J.-L., L. Marivaux, P.-O. Antoine, and M. Benammi. 1999. Paleontologie dans les Bugti Hills, Nouvelle donnees. *Bulletin de la Societe d'Histoire Naturelle de Toulouse* 135:135–139.

Welcomme, J. L., P.-O. Antoine, F. Duranthon, P. Mein, and L. Ginsburg. 1997. Nouvelles découvertes de Vertébrés miocènes dans le synclinal de Dera Bugti (Balouchistan, Pakistan). *Comptes rendus de l'Académie des sciences. Série 2. Sciences de la terre et des planètes* A 325(7):531–536.

Welcomme, J.-L., M. Benammi, J.-Y. Crochet, L. Marivaux, G. Metais, P.-O. Antoine, and I. Baloch. 2001. Himalayan Forelands: Paleontological evidence for Oligocene detrital deposits in the Bugti Hills (Balochistan, Pakistan). *Geological Magazine,* 138(4):397–405.

Wolfe, J. A. 1971. Tertiary climatic fluctuations and methods of analysis of Tertiary floras. *Palaeogeography, Palaeoclimatology, Palaeoecology* 9:27–57.

Wolfe, J. A. 1978. A paleobotanical interpretation of Tertiary climates in the Northern Hemisphere. *American Scientist* 66:694–703.

Wolfe, J. A. 1980. Tertiary climates and floristic relationships at high latitudes in the Northern Hemisphere. Palaeogeography, Palaeoclimatology, Palaeoecology 30:313–323.

Wolfe, J. A. 1985. Distributions of major vegetational types during the Tertiary. *American Geophysical Union Geophysical Monographs* 32:357–376.

Wolfe, J. A. 1992. Climatic, floristic, and vegetational changes near the Eocene/Oligocene boundary in North America, pp. 421–436, *in* Prothero, D. R., and W. A. Berggren (eds.), *Eocene-Oligocene Climatic and Biotic Evolution*. Princeton University Press, Princeton, New Jersey.

Wood, H. E., II. 1938. *Cooperia totadentata,* a remarkable rhinoceros from the Eocene of Mongolia. *American Museum Novitates* 1012:1–20.

Woodburne, M. O. (ed.), 2004. *Late Cretaceous and Cenozoic Mammals of North America.* Columbia University Press, New York.

Xu, Y., and J. Wang. 1978. New materials of giant rhinoceros. *Memoirs of the Institute of Vertebrate Paleontology and Paleoanthropology* 13:132–140.

Yablokov, A. V. 1974. *Variability in Mammals.* Amerind Press, New Delhi.

Yassamanov, N. Y. 1985. *Ancient Climates.* Hydrometeoizdat, Leningrad.

Stalin, 24
Stegosaurus, 7
Stehlin, Hans, 113
Stenoplesictis, 39
Stylemys nebraskensis, 110
Sunketanka, 71
Svita, 47-50
Swisher, Carl C., 41
Syntype, 77
Systematics, 67-69

Taben Buluk, 42
Tabenbulukian, 42, 50, 119
Taiwan Geological Survey, 31
Taliban, 34
Tapir, 54
Taxidermy, 4, 7
Taxon, 67-69
Taxonomy, 67
Teilhard de Chardin, Pierre, 28-30
Thaumastotherium, 23, 74-75, 77
Thermoregulation, 100-101
Thighbone, 94-95
Tortoises, 110
Trilophodon cooperi, 24
Tring, Hertfordshire, 23
Trinity College, Cambridge, 21
Triplopodinae, 76
Triplopus, 60-61
Trucial Oman, 18
Tsintaosaurus, 30
Turgai, 26
Turkey, 19, 34, 50-51

Turpanotherium, 78, 80-81
Type specimen, 76-77
Tyrannosaurus, 9

Uintan, 42
University College, London, 17
University Museum of Zoology, Cambridge, 23
Urtinotherium, 31, 41, 63-66, 80-81, 87
Urtyn Obo, 13
USS Albatross, 4-5

Velociraptor, 9
Vertebrae, 91-92

Weidenreich, Franz von, 29
Welcomme, Jean-Loup, 34-38, 83
Whitneyan, 42
Willamette flora, 111
Wood, Horace E. II, 24, 74
World War I, 19, 23
World War II, 23-24

Yang Zhong-Jian, 30-31
Yugoslavia, 51
Yunnan Province, 5
Yunnanosaurus, 30

Zaisanamynodon, 48
Zaysan Basin, 47-50
Zdansky, Otto, 28
Zhou Ming-Zhen, 30-32
Zhoukoudian caves, 7, 28, 30

DONALD R. PROTHERO has taught college geology and paleontology for 33 years at institutions including the California Institute of Technology and Columbia, Pierce, Occidental, Knox, and Vassar Colleges. He is the author of more that 35 books (including five geology textbooks) and over 250 scientific papers. He is a Fellow of the Geological Society of America, the Paleontological Society, the Linnean Society of London, and in 1988 he was a Guggenheim Fellow. In 1991, he received the Charles Schuchert Award of the Paleontological Society for the outstanding paleontologist under the age of 40.